Edexcel A Level

CW00520783

MATHEMATICS

EXAM PRACTICE

For Year 1 and AS

1

HODDER
EDUCATION
AN HACHETTE UK COMPANY

Hachette UK's policy is to use papers that are natural, renewable and recyclable products and made from wood grown in sustainable forests. The logging and manufacturing processes are expected to conform to the environmental regulations of the country of origin.

Orders: please contact Bookpoint Ltd, 130 Park Drive, Milton Park, Abingdon, Oxon OX14 4SE. Telephone: (44) 01235 827720. Fax: (44) 01235 400454.

Email education@bookpoint.co.uk Lines are open from 9 a.m. to 5 p.m., Monday to Saturday, with a 24-hour message answering service. You can also order through our website: www.hoddereducation.co.uk

ISBN: 978 1 5104 2363 3

© Heather Davis, Janet Dangerfield, Nick Geere, Rose Jewell, Sue Pope, Andrew Roberts 2018

First published in 2018 by

Hodder Education,

An Hachette UK Company

Carmelite House

50 Victoria Embankment

London EC4Y 0DZ

www.hoddereducation.co.uk

Impression number 10 9 8 7 6 5 4 3 2 1

Year 2022 2021 2020 2019 2018

Cover photo © Baloncici/123RF.com

Typeset in Integra Software Services Pvt. Ltd., Pondicherry, India

Printed in the UK by CPI Group Ltd

A catalogue record for this title is available from the British Library.

MIX
Paper from
responsible sources
FSC™ C104740
FSC
www.fsc.org

Contents

Introduction   iv

1   Proof   1

2   Surds and indices   3

3   Quadratic functions   4

4   Equations and inequalities   5

5   Coordinate geometry   6

6   Trigonometry   8

7   Polynomials   10

8   Graphs and transformations   11

9   The binomial expansion   14

10   Differentiation   15

11   Integration   18

12   Vectors   19

13   Exponentials and logarithms   21

14   Data collection   24

15   Data processing, presentation and interpretation   25

16   Probability   32

17   The binomial distribution   34

18   Statistical hypothesis testing using the binomial distribution   36

19   Kinematics   38

20   Forces and Newton's laws of motion   42

21   Variable acceleration   45

Answers   47

**Full worked solutions and mark schemes are available at
www.hoddereducation.co.uk/EdexcelMathsExamPractice**

Introduction

This book offers over 300 questions to support successful preparation for the new AS and A levels in Mathematics. Grouped according to topic, the chapters follow the content of *Edexcel A level Mathematics Year 1 (AS)*. Each chapter starts with short questions to support retrieval of content and straightforward application of skills learned during the course. The demand gradually builds through each chapter, with the later questions requiring significant mathematical thinking and including connections to other topics. This reflects the likely range of styles of question in the live papers. Answers are provided in this book, and worked solutions and mark schemes to all questions can be found online at **www.hoddereducation.co.uk/EdexcelMathsExamPractice.**

The data set referred to in some of the questions on statistics can be found at www.ocr.org.uk/qualifications/as-a-level-gce/as-a-level-gce-mathematics-b-mei-h630-h640-from-2017/assessment/

1 Proof

1 Given that n is the smallest of three consecutive integers, write
   down expressions for the other two numbers. [1 mark]

2 If n is a positive integer, which of the following expressions is
   never even?
   $n + 4$ $3n - 5$ $2n + 2$ $n + 2$ $2n + 1$ [1 mark]

3 P: X has three factors.
   Q: X is a square number.
   Which of the following is true?
   $P \Rightarrow Q$ $P \Leftarrow Q$ $P \Leftrightarrow Q$ [1 mark]

4 n is any positive integer.
   Prove by counter example that $n^2 + 3n + 13$ is not always prime. [2 marks]

5 'No cube numbers end in 7.'
   Prove by counter example that this is not true. [1 mark]

6 Prove that the square of an odd number is also odd. [3 marks]

7 (i) Prove that the product of three consecutive
        integers is divisible by six. [3 marks]

   (ii) When is the product of three consecutive
        integers a multiple of 12?
        Justify your answer. [2 marks]

8 (i) Prove that the equation $x^2 = mx + c$ has two
        distinct real roots when $m^2 + 4c > 0$. [4 marks]

   (ii) Find an expression for c, such that $y = mx + c$ is a
        tangent to the graph of $y = x^2$. [2 marks]

9 An equable shape is one whose perimeter is numerically equal to its area.

   (i) For equable rectangles with length and width a and b respectively,
        find an expression for a in terms of b. [3 marks]

   (ii) Give two examples to show why equable
        rectangles do not exist for $b \le 2$. [2 marks]

10 'A pair of simultaneous linear equations
   $ax + by = c$
   $dx + ey = f$
   always has a single solution.'
   Under what conditions is this statement true?
   Justify your answer. [6 marks]

11 (i) A piece of A-size paper is folded in half.
It forms a rectangle that is similar to the original piece of paper.
Show that the ratio of the sides of A-size paper is $1 : \sqrt{2}$,
explaining your method carefully. [4 marks]

(ii) A piece of A-size paper is folded so that the short edge meets
the long edge and then the new shortest edge meets the original
short edge, as shown in the diagram below.
Prove that the shape formed is a kite. [4 marks]

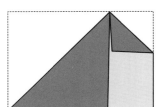 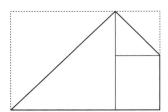

2 Surds and indices

1 Express $\sqrt{75}$ in the form $a\sqrt{b}$, where a and b are integers > 1. [1 mark]

2 Simplify $\sqrt{72} - \sqrt{50}$. [2 marks]

3 Rationalise the denominator of $\dfrac{5}{\sqrt{7}}$. [1 mark]

4 Find the exact value of $\left(2\tfrac{7}{9}\right)^{\frac{1}{2}}$. [2 marks]

5 Find the exact value of $\left(3\tfrac{3}{8}\right)^{-1}$. [2 marks]

6 Simplify $\left(\sqrt{7} - \sqrt{2}\right)\left(2\sqrt{7} + 7\sqrt{2}\right)$. [2 marks]

7 Express $\dfrac{2+\sqrt{2}}{2-\sqrt{2}}$ in the form $a + b\sqrt{2}$, where a and b are integers. [2 marks]

8 Simplify:

(i) $\left(\dfrac{1}{4}\right)^{-\frac{1}{2}}$ [1 mark]

(ii) $\left(\dfrac{4}{5}\right)^{0}$ [1 mark]

(iii) $\left(\dfrac{343}{125}\right)^{\frac{2}{3}}$ [2 marks]

9 Simplify $\dfrac{(5ab^2)^3 \times (2a^3b)^2}{(10a^3b^2)^3}$. [2 marks]

10 Simplify $\left(8a^{\frac{7}{2}}\right)^{\frac{1}{3}} \times \left(16a^{\frac{2}{3}}\right)^{-\frac{1}{4}}$. [2 marks]

11 The triangle below is made up of two right-angled triangles. One is half an equilateral triangle with side length $2x$.

Show that $\sin\theta = \dfrac{\sqrt{21}}{7}$, showing your method clearly. [6 marks]

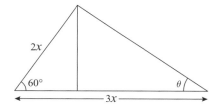

12 (i) A line segment of unit length is split into two parts such that the ratio of the whole to the longest part is equal to the ratio of the longer part to the shorter part.

Given that the longer part is x, establish that: $\dfrac{1}{x} = \dfrac{x}{1-x}$. [2 marks]

(ii) Solve the equation $\dfrac{1}{x} = \dfrac{x}{1-x}$ for x. [4 marks]

(iii) The Golden Ratio is a number that has the property that its reciprocal is equal to one more than the Golden Ratio. Show that x has this property. [3 marks]

3 Quadratic functions

1. Determine the exact solutions of $x^2 - x - 1 = 0$. [2 marks]

2. Factorise $4x^2 - y^2$. [1 mark]

3. Solve $3x^2 - 2x - 2 = 0$.
   Give your answers correct to 3 decimal places. [2 marks]

4. Express $x^2 + 6x - 5$ in the form $(x + a)^2 + b$ where a and b are integers. [2 marks]

5. Solve the equation $4x^4 - 5x^2 - 9 = 0$. [3 marks]

6. Write down the coordinates of the turning point of the curve $y = 3(x - 1)^2 + 5$. [1 mark]

7. Given that the equation $x^2 - kx + 1 = 0$ has one distinct root, find the set of possible values of k. [2 marks]

8. Write down the equation of the line of symmetry of the curve $y = 6(x + 3)^2 - 1$. [1 mark]

9. (i) Find the minimum point of the function $y = 2x^2 - x - 3$. [3 marks]

   (ii) Sketch the graph of $y = 2x^2 - x - 3$. [3 marks]

10. The line $y = kx + 2$ intersects the curve $y = 2x^2 - 3x + 5$ twice. Determine, in exact form, the values of k for which this occurs. [6 marks]

11. (i) Factorise the expression $10x^2 - 7x - 12$. [2 marks]

    (ii) Solve the equation $10x^2 - 7x - 12 = 0$. [1 mark]

    (iii) Solve the equation $10x^4 - 7x^2 - 12 = 0$, giving your answers in the form $x = a\sqrt{b}$, where a is a rational number and b is an integer. [2 marks]

    (iv) Solve the inequality $10x^2 - 7x - 12 > 0$ algebraically, and illustrate the solution graphically. [4 marks]

12. (i) Determine the equation of the function that has the gradient function given by:

    $$\frac{dy}{dx} = 4x + 1$$

    and that passes through the point with coordinates (1,2). [4 marks]

    (ii) Sketch the graph of this function, identifying the points where the graph crosses the axes and also any turning points. [5 marks]

4 Equations and inequalities

1. x is an integer.
   List the members of the set $\{x : x \geqslant -1\} \cap \{x : x < 5\}$. [1 mark]

2. Describe the values of x that are not included in $x < -6$ or $x \geqslant -2$. [1 mark]

3. Solve the following simultaneous equations.
   $2x - 5y = 15$
   $3y + 4x = 4$ [3 marks]

4. Represent the region $2x + y < 3$ graphically. [2 marks]

5. Solve the following simultaneous equations.
   $x^2 + y^2 = 10$
   $y - 2x + 7 = 0$ [4 marks]

6. Solve the inequality $x^2 > 9$. [2 marks]

7. Solve the inequality $6 + x - x^2 \geqslant 0$. [3 marks]

8. Find the point of intersection of the lines $2x - 3y + 5 = 0$
   and $7x + 2y - 4 = 0$. [3 marks]

9. Solve the inequality $-2 - 5x > -3 - 2x$. [2 marks]

10. A car journey of 100 miles takes 2 hours 30 minutes.
    Part of it is spent on the motorway, travelling at 70 m.p.h., and
    the rest of it is spent on country roads, travelling at 30 m.p.h.

    By writing this information as a pair of simultaneous
    equations, find the distances travelled on each type of road. [4 marks]

11. Find the coordinates of the points where the line $x - y = 4$
    meets the curve $4x^2 + 7y^2 = 67$. [5 marks]

12. Solve the inequality $a(a - 2) > 3$, expressing your solution
    using set notation and the symbols $\cup$ and $\cap$. [4 marks]

5 Coordinate geometry

1 Find the gradient of the line joining the points $(-2, 1)$ and $(4, -3)$. [1 mark]

2 Write down the coordinates of the midpoint of the line joining the points $(-5, 2)$ and $(1, 7)$. [1 mark]

3 Work out the distance between the points $(-5, -1)$ and $(7, 4)$. [1 mark]

4 Write down the gradient of the line perpendicular to the line with equation $y = \frac{1}{3}x - 3$. [1 mark]

5 Find the equation of the line joining the points $(-2, -2)$ and $(1, 4)$. [2 marks]

6 A circle has the equation $(x + 4)^2 + (y - 1)^2 = 7$.
Write down the coordinates of its centre and its radius. [2 marks]

7 Determine the coordinates of the points where the line $2x + 5y = 10$ crosses the x and y axes. [2 marks]

8 Show that the point $(1, 1)$ lies below the line with equation $y = 5 - 3x$. [2 marks]

9 Show that the point $(3, -6)$ lies inside the circle with equation $(x - 1)^2 + (y + 4)^2 = 9$. [2 marks]

10 Find the coordinates of the point of intersection of the lines $y + 2x = 3$ and $y = 5 + 7x$. [3 marks]

11 Find the gradient of the tangent to the circle with equation $x^2 + y^2 = 5$ at the point $(-2, 1)$. [2 marks]

12 Find the coordinates of the points of intersection of the curves with equations $y = 3x^2 - 2$ and $y = 5 - x^2$. [4 marks]

13 Show that the line with equation $x + y = 8$ does not intersect the circle with equation $(x + 3)^2 + y^2 = 9$. [4 marks]

14 Find the equation of the tangent to the circle with equation $x^2 + y^2 = 13$ at the point $(-2, -3)$. [4 marks]

15 The line l passes through the point A$(4, 4)$ and has gradient 2.

  (i) Find the equation of line l in the form $ax + by + c = 0$, where a, b and c are integers. [2 marks]

  (ii) Show that point B$(1, -2)$ also lies on line l. [1 mark]

  (iii) Find, in exact form, the distance AB. [2 marks]

  C is the point $(1, 3)$.

  (iv) Work out the area of the triangle ABC. [2 marks]

  (v) Hence, or otherwise, find the shortest distance from C to the line l. [2 marks]

16 A circle has the centre $(5, 3)$, and the y axis is a tangent to the circle.

  (i) Write down the radius of the circle. [1 mark]

  (ii) Write down the equation of the circle. [2 marks]

  (iii) Show that the circle intersects the x axis at $(1, 0)$. [2 marks]

  (iv) Find the points on the circle where the tangent is parallel to the diameter through $(1, 0)$. [4 marks]

17 The line l is the perpendicular bisector of the points A$(1, 8)$ and B$(5, 6)$.

    (i) Find the equation of the line l in the form $y - y_1 = m(x - x_1)$. [4 marks]

    The line L has the equation $4x - 3y = 30$.

    (ii) Sketch the line L on coordinate axes, giving the coordinates of the points where the line L cuts the coordinate axes. [3 marks]

    (iii) Work out the coordinates of the point C, where line l meets line L. [3 marks]

18 A circle passes through points A$(1,1)$, B$(4,k)$, C$(6,k)$ and D$(11,11)$, for some constant value k.

    (i) Find the equation of the perpendicular bisector of AD, giving your answer in the form

    $ax + by + c = 0$,

    where a, b and c are integers. [5 marks]

    (ii) Explain why the x-coordinate of the centre of the circle must be 5. [1 mark]

    (iii) Work out the coordinates of the centre of the circle. [2 marks]

    (iv) Find the equation of the circle. [3 marks]

    (v) Hence, find the possible values of k. [3 marks]

19 (i) Work out the coordinates of the points where the line

    $2x + y - 5 = 0$

    cuts the curve

    $y^2 = 4x + 5$. [4 marks]

    (ii) Work out the coordinates of the points where the line

    $2x + y - 5 = 0$

    cuts the circle

    $x^2 + y^2 - 10x - 6y = 30$. [4 marks]

    (iii) Hence, find the coordinates of one of the points where the curve

    $y^2 = 4x + 5$

    cuts the circle

    $x^2 + y^2 - 10x - 6y = 30$ [2 marks]

20 Points A$(-2, 3)$ and B$(10, 8)$ are the points at each end of the diameter of a circle.

Point C lies on the circle and is nearer to B than it is to A.

Point D lies on the circle on the opposite side of the diameter AB to point C.

ACBD is a kite shape of area 78 units^2.

Find the exact coordinates of the point E, where the chord CD cuts the diameter AB. [9 marks]

6 Trigonometry

1 The diagram shows the unit circle.

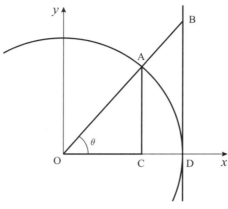

    (i) What length on the diagram represents $\cos \theta$?   [1 mark]

    (ii) What length on the diagram represents $\tan \theta$?   [1 mark]

2 Work out the value of θ in degrees.   [2 marks]

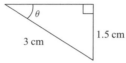

3 Solve the equation $\cos \theta = -0.3$ for $0° \leqslant \theta \leqslant 180°$.   [1 mark]

4 Solve the equation $\sin \theta = 0.7$ for $90° \leqslant \theta \leqslant 180°$.   [2 marks]

5 Solve the equation $\tan \theta = 5$ for $180° \leqslant \theta \leqslant 360°$.   [2 marks]

6 Work out the value of θ in degrees.   [3 marks]

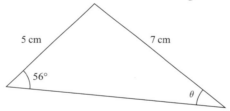

7 Work out the value of θ in degrees.   [3 marks]

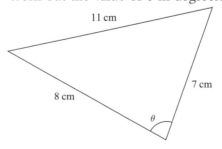

8 Work out the area of the triangle.   [2 marks]

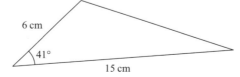

9 Find the height, in exact form, of an equilateral triangle with side length 15 cm.   [2 marks]

10 Work out h for the regular octagon with side length 5 cm.　　[3 marks]

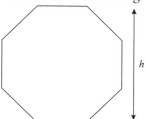

11 Solve the equation $\sqrt{5}\sin\theta = 2\cos\theta$ for $0° \leqslant \theta \leqslant 180°$.　　[2 marks]

12 Find the values of θ in the range $-180° < \theta < 180°$ that satisfy the equation $\sin 2\theta = 0.5$.　　[4 marks]

13 Find the values of θ in the range $0° < \theta < 1080°$ that satisfy the equation $\cos\left(\dfrac{\theta}{2}\right) = \dfrac{\sqrt{3}}{2}$.　　[4 marks]

14 Find the values of θ in the range $0° \leqslant \theta \leqslant 360°$ that satisfy the equation $\sin^4\theta + \cos^2\theta = \dfrac{13}{16}$.　　[6 marks]

15 In the triangle ABC, AB is of length 12 cm, BC is of length 18 cm and the angle at C is 40°.
Find the two possible areas of the triangle.　　[6 marks]

16

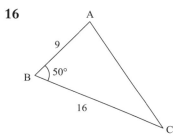

The diagram shows the relative positions of three ships at sea.
A is 9 nautical miles from B, and C is 16 nautical miles from B.
A is on a bearing of 060° from B.
Find the bearing of C from A.　　[6 marks]

17 In this question you must show detailed reasoning.

Simplify the following expressions, giving your answers in terms of $\sin\theta$ and $\cos\theta$:

(i) $\sin(\theta - 90°) + \sin(\theta + 90°)$　　[2 marks]

(ii) $\cos(-\theta) + \cos(90° - \theta) + \cos(180° - \theta)$　　[3 marks]

(iii) $\sin(180° + \theta) + \sin(180° - \theta) + \cos(180° + \theta) + \cos(180° - \theta)$ [4 marks]

18 Show that $\dfrac{1 + \cos\theta - \sin\theta}{1 + \cos\theta + \sin\theta} = \dfrac{1}{\cos\theta} - \tan\theta$　　[5 marks]

7 Polynomials

1. $f(x) = 2x^3 - x^2 + 5x - 2$
   $g(x) = 5x^3 - 2x^2 - 5$
   Work out $g(x) - f(x)$. [1 mark]

2. Expand and simplify $(x - 5)(2x^2 - x + 3)$. [2 marks]

3. A polynomial of degree m is multiplied by a polynomial of degree n.
   State the degree of the product. [1 mark]

4. A polynomial of degree m is divided by a polynomial of degree n.
   State the degree of the quotient. [1 mark]

5. A polynomial of degree 3 has p distinct roots.
   Which of the following cannot be the value of p?
   0   1   2   3 [1 mark]

6. Divide the polynomial $f(x) = 6x^3 - 11x^2 + 11x - 12$
   by $(2x - 3)$. [3 marks]

7. **(i)** Show that $x - 2$ is a factor of the polynomial
   $f(x) = 2x^3 + 7x^2 - 10x - 24$. [2 marks]

   (ii) Factorise $f(x)$ completely. [4 marks]

   (iii) Sketch $y = f(x)$. [3 marks]

8. Determine whether there is an integer root of the equation
   $x^3 + 2x^2 - 5x + 3 = 0$. [5 marks]

9. **(i)** Given that $x - 2$ and $x + 3$ are factors of the polynomial
   $f(x) = x^3 + ax^2 + bx + 30$, find a and b. [6 marks]

   (ii) Hence, sketch the curve $y = f(x)$. [3 marks]

10. It is given that $f(x) = (x + 3)(x - 1)(x - 4)$.

    (i) Sketch the curve $y = f(x)$. [3 marks]

    (ii) Show that $f(x)$ may be written as $x^3 - 2x^2 - 11x + 12$. [2 marks]

    The function, $g(x) = f(x + 1)$

    (iii) State the roots of $g(x) = 0$. [2 marks]

11. A cubic curve intersects the axes at the points $(-3, 0)$, $(1, 0)$,
    $(2.5, 0)$ and $(0, 30)$.

    (i) Express the equation of the curve in factorised form. [2 marks]

    (ii) Show that the equation of the curve may be written as
    $y = 4x^3 - 2x^2 - 32x + 30$. [2 marks]

12. It is given that $f(x) = (x - 2)(x + a)^2$, where a is a positive integer.

    (i) Sketch the graph of $y = f(x)$. [3 marks]

    The curve crosses the y axis at $(0, -50)$.

    (ii) Determine the value of a. [2 marks]

8 Graphs and transformations

1 Describe the transformation that maps the curve $f(x) = x^2$
   onto the curve $g(x) = -x^2$. [1 mark]

2 The graph of $y = x^2$ is translated by the vector $\begin{pmatrix} 3 \\ 0 \end{pmatrix}$.

   Write down the equation of the resulting graph. [1 mark]

3 The graph of $y = \cos x$ is transformed to the graph of $y = 2\cos x$.
   Describe the transformation. [1 mark]

4 The graph of $x^2 + y^2 = 5$ is translated by the vector $\begin{pmatrix} -2 \\ 5 \end{pmatrix}$.
   Write down the equation of the resulting graph. [2 marks]

5 The graph of $y = e^x$ is reflected in the y axis.
   Write down the equation of the resulting graph. [1 mark]

6 Given that $\sin x = \cos(x - 90°)$, describe a transformation
   that maps the graph of $y = \cos x$ onto the graph of $y = \sin x$. [1 mark]

7 The graph of $y = x^3 - x^2 + 1$ is translated by the vector $\begin{pmatrix} 1 \\ -2 \end{pmatrix}$.
   Write down the equation of the resulting graph. [2 marks]

8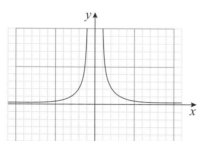

   This graph represents one of the following functions.
   Which one?

   $y = -\dfrac{1}{x}$

   $y = x^3$

   $y = \dfrac{1}{x^3}$

   $y = \dfrac{1}{x^2}$ [1 mark]

9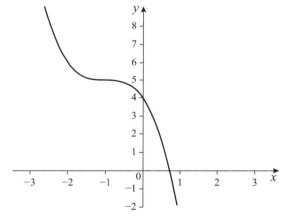

Which one of the following equations is represented by this graph?

$y = -x^3 - x^2 + 12x$

$y = x^4 + x^3 - 6x^2$

$y = x^3 + x^2 - 6x$

$y = x^2 + x - 6$ [1 mark]

10 Sketch a graph that illustrates direct proportion.   [1 mark]

11 Sketch a graph that illustrates indirect proportion.   [1 mark]

12 **(i)** Sketch the graph of $y = \dfrac{x+1}{x-1}$, labelling intercepts
        and asymptotes clearly.   [5 marks]

   (ii) Describe the transformation that maps $y = \dfrac{1}{x}$ onto $y = \dfrac{x+1}{x-1}$. [2 marks]

13 A polynomial has roots $-3, -1, 2$ and 5 and a y-intercept of 3.
   Sketch the graph, and find the equation of the function.   [6 marks]

14 A cubic function is reflected in the x axis and translated by

   the vector $\begin{pmatrix} -1 \\ 5 \end{pmatrix}$.

   The following graph is obtained.

   What was the original function?   [4 marks]

15 Describe the transformation that maps the curve $y = e^x$ to the curve:

   (i) $y = e^x + 3$ [1 mark]

   (ii) $y = 3e^x$ [1 mark]

   (iii) $y = e^{3x}$ [1 mark]

   (iv) $y = e^{x+3}$ [1 mark]

Describe a sequence of transformations that map the curve $y = e^x$ to the curve:

(v) $y = 3e^{3x+3} + 3$ [3 marks]

16 For the table of values below, the function that relates them is of the form $y = a\sqrt{x} + b$.

By plotting a suitable graph, or otherwise, work out the values of a and b. [5 marks]

x	1.0	1.5	2.0	2.5	3.0
y	5.5	6.4	7.2	7.8	8.4

17 The London Eye is a big wheel with a radius of 60 m and a maximum height of 135 m.

It takes 30 minutes for a single rotation.

Determine the function that gives the height above the ground (h) in metres, as a function of time (t) in minutes. Assume the ride starts at its lowest point. [6 marks]

18 Complete the table below.
The first row has been done for you. [6 marks]

Function	Period	Coordinates of maximum closest to (0,0)	Coordinates of minimum closest to (0,0)
$y = 5\sin 2x + 1$	180°	(45°, 6)	(−45°, −4)
$y = 3\cos\left(\frac{x}{2} - 30°\right) + 5$			
		(22.5°, 3)	(−22.5°, −1)

9 The binomial expansion

1 Simplify $\frac{6!}{4!}$ [1 mark]

2 Simplify $_5C_3$. [1 mark]

3 Find the constant term in the binomial expansion of $(3 - 2x)^6$. [1 mark]

4 Find the coefficient of x in the binomial expansion of $(1 + 3x)^6$. [1 mark]

5 Find the coefficient of x^2 in the binomial expansion of $(1 - 2x)^7$. [2 marks]

6 Find the coefficient of x^2 in the binomial expansion of $(3 + 2x)^5$. [2 marks]

7 **(i)** Write down the first four terms of the binomial expansion of $(2 - 3x)^5$ (in increasing powers of x), in simplified form. [5 marks]

   (ii) Hence, obtain an approximation for $(1.97)^5$ to 5 significant figures. [3 marks]

8 Expand $(1 + 2x)(1 - 3x)^3$. [4 marks]

9 A coin is tossed eight times. How many of the possible outcomes include exactly three heads? [2 marks]

10 A House of Commons committee is to consist of 12 members, of which 6 are to be Conservative, 4 Labour, and 2 from other parties. 10 Conservative MPs have put their names forward, along with 8 Labour and 6 from other parties. The committee members are chosen at random. What is the probability of a given committee being selected? [5 marks]

10 Differentiation

1 A function, $y = f(x)$, is differentiated with respect to x.
What is the significance of the function obtained? [1 mark]

2 The gradient function for a graph is given by $\dfrac{dy}{dx} = 2x + 3$.
Find the gradient of the graph at the point where $x = 4$. [1 mark]

3 State the gradient function for $y = 7x - 1$. [1 mark]

4 State the gradient function for $y = x^3$. [1 mark]

5 Given that $y = x^2 - 3x$, find $\dfrac{dy}{dx}$. [2 marks]

6 Given that $f(x) = 2x^3 - 4x + 7$, find $f'(x)$ [2 marks]

7 Given that $y = \dfrac{1}{x^2}$, find $\dfrac{dy}{dx}$. [2 marks]

8 Given that $y = \sqrt{x}$, find $\dfrac{dy}{dx}$. [2 marks]

9 Find the derivative, with respect to x, of the function
$y = x^5 - x + 1$. [2 marks]

10 Sketch the gradient function for the following graph. [3 marks]

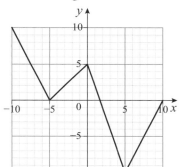

11 Sketch the velocity–time graph for the journey shown in the
following distance travelled–time graph. [3 marks]

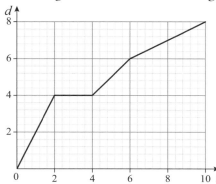

12 The function $y = -x^3 - 6x^2 - 9x + 1$ is an increasing function
for certain values of x.

Determine those values. [4 marks]

13 For what values of x is the function $y = x^3 + x^2 - 5x - 3$ a
decreasing function? [4 marks]

14 (i) Work out the gradient of the chord joining $(1,1)$ to $(2,8)$ on the graph of $y = x^3$. [1 mark]

(ii) Work out the gradient of the chord joining $(1,1)$ to $(1.1, 1.1^3)$ on the graph of $y = x^3$. [1 mark]

(iii) Work out the gradient of the chord joining $(1,1)$ to $(1.01, 1.01^3)$ on the graph of $y = x^3$. [1 mark]

(iv) What can you deduce about the gradient of the tangent to the curve at $(1,1)$? [1 mark]

15 The first derivative of a function is zero at a particular point, and the second derivative is positive at that point. What can you deduce about the function at that point? [1 mark]

16 Differentiate the function $y = x^3$ from first principles, with respect to x. [3 marks]

17 Differentiate the function $y = x^2 - 4x + 5$ from first principles, with respect to x. [3 marks]

18 The spreadsheet shows the gradient of the chord joining $(2,2)$ and points to the right of it on the graph of $y = x^3 - 3x$ with an x increment of $0.1, 0.01, 0.001$, and so on.

	A	B	C
	x	**x^3-3x**	**gradient**
1			
2	2	2	
3	2.1	2.961	9.61
4	2.01	2.090601	9.0601
5	2.001	2.009006	9.006001
6	2.0001	2.0009	9.0006
7	2.00001	2.00009	9.00006

Write down the value that the gradient is approaching, and verify that it is the gradient of the curve at $(2,2)$ using differentiation. [3 marks]

19 The gradient of the tangent to the curve $y = x^2 + ax + b$ at the point $(-1, 5)$ is 4.

Work out the values of a and b. [4 marks]

20 (i) Show that the graphs of $y = x^3 + x$ and $y = x$ have one point in common. [2 marks]

(ii) Show that the gradients of their graphs are equal at that point. [3 marks]

(iii) What must be true about the relationship between these functions at that point? [1 mark]

21 The cost C of a ship's journey is given by $C = 100v^2 + \dfrac{200\,000}{v}$, where v is the ship's speed in knots.

(i) Find $\dfrac{\mathrm{d}C}{\mathrm{d}v}$. [2 marks]

(ii) For what values of v is the cost C increasing? [3 marks]

(iii) What is the most economical speed? [3 marks]

22 (i) On the curve $y = x + \dfrac{1}{x}$, find the points, A and B, at which the tangents are parallel to the x axis, and write down the equations of the tangents. [5 marks]

(ii) Write down the equations of the normals to the curve at points A and B. [1 mark]

(iii) The tangents and normals enclose a rectangle with the origin at its centre.

Find the area of the rectangle. [2 marks]

23 (i) Find the equations of the tangent and the normal to the curve $y = (x-1)(x^2+2)$ at the point P(2,6). [8 marks]

(ii) The tangent cuts the x axis at A, and the normal cuts the x axis at B. Find the area of triangle ABP. [4 marks]

24 (i) A cuboid with a square base has a total surface area of $150\,\text{cm}^2$. If the side of the base is $x\,\text{cm}$, show that the volume of the cuboid is $\frac{x}{2}(75 - x^2)$. [5 marks]

(ii) Find the maximum volume of the cuboid. [5 marks]

25 An 18 cm length of wire is formed into a pentagon consisting of a rectangle ABCD and an equilateral triangle ADE, with side length x.

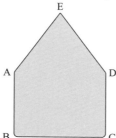

(i) Show that the area can be expressed as $9x + x^2\left(\frac{\sqrt{3}}{4} - \frac{3}{2}\right)$. [6 marks]

(ii) Determine the maximum area of the pentagon, to 3 significant figures. [7 marks]

(iii) Determine, in exact form, the length of the side of the triangle and the height of the rectangle. [4 marks]

26 A cylindrical metal can holds 500 ml.
Assuming the metal is of uniform thickness and ignoring any additional material needed for the joins, determine the minimum amount of metal that can be used to make the can. Give your answer to an accuracy of 3 significant figures. [11 marks]

11 Integration

1. Given that $\frac{dy}{dx} = 2x$, find the function y. [1 mark]

2. Given that $\frac{dy}{dx} = 3x^2 - 2x$, find the function y. [2 marks]

3. Given that $f'(x) = \sqrt{x}$, find the function $f(x)$. [2 marks]

4. A curve has gradient function given by $\frac{dy}{dx} = 5x - 2$ and passes through the point $(-2, 9)$.
   Find the equation of the function. [4 marks]

5. A function, y, has gradient function $\frac{dy}{dx} = \frac{4}{x^2} + 3$, and $y = 1$ when $x = 2$.
   Find the equation of the function. [4 marks]

6. Find $\int_1^4 (3\sqrt{x} + 2)\, dx$. [4 marks]

7. Find $\int x\sqrt{x}\, dx$. [2 marks]

8. The velocity of a particle, at time t, is given by $v = 6t^2 - 4t$. Express s, the displacement of the particle at time t, as a function of t. [2 marks]

9. Find $\int_{-1}^3 (x^2 - 2x + 4)\, dx$. [4 marks]

10. Find $\int (x^3 - \frac{1}{x^3})\, dx$. [3 marks]

11. Find the area of the region bounded by the curve $y = \frac{4}{\sqrt{x}}$, the lines $x = 4$ and $x = 9$, and the x axis. [4 marks]

12. The gradient of a curve is given by $8x(x^2 - 1)$, and the curve passes through the point $(2, 10)$.
    Find the coordinates of the stationary points of the curve. [5 marks]

13. Find the total area of the two regions between the curve $y = (x - 1)(x + 2)(x + 3)$ and the x axis. [5 marks]

14. Find $\int_1^2 (x + 2)(x + 3)\, dx$. [6 marks]

15. (i) On the same diagram, sketch graphs of $y = (x + 2)^2 + 1$ and $y = 3 - x$. [3 marks]
    (ii) Find the area of the region satisfying both $y > (x + 2)^2 + 1$ and $y < 3 - x$, giving your answer to 3 significant figures [10 marks]

12 Vectors

1 Given that $\mathbf{a} = \begin{pmatrix} 3 \\ -4 \end{pmatrix}$ and $\mathbf{b} = \begin{pmatrix} -1 \\ 2 \end{pmatrix}$, work out $2\mathbf{a} - \mathbf{b}$. [1 mark]

2 Given that $\mathbf{p} = 8\mathbf{i} - 15\mathbf{j}$, what is the magnitude of $\mathbf{p}$? [1 mark]

3 $\mathbf{i}$ and $\mathbf{j}$ are unit vectors in the east and north directions, respectively.
   Find the bearing of the vector $2\mathbf{i} - 5\mathbf{j}$. [1 mark]

4 A vector has magnitude 12 and makes an angle of $-30°$ with the positive x axis.
   Write the vector in component form. [2 marks]

5 Write down the unit vector parallel to $-2\mathbf{i} + \mathbf{j}$. [2 marks]

6 Point A has the position vector $-3\mathbf{i} + 2\mathbf{j}$.
   Point B has the position vector $\mathbf{i} - 4\mathbf{j}$.
   Work out the length of AB. [2 marks]

7 The vertices of the triangle ABC have the position vectors $\mathbf{a} = \mathbf{i} + \mathbf{j}$, $\mathbf{b} = -3\mathbf{i} + 4\mathbf{j}$ and $\mathbf{c} = 4\mathbf{i} + 5\mathbf{j}$, respectively.
   (i)   Write down the vector $\overrightarrow{AB}$ in terms of $\mathbf{i}$ and $\mathbf{j}$. [2 marks]
   (ii) Find the unit vector in the same direction as $\overrightarrow{AB}$. [2 marks]
   (iii) Use a vector method to show that the triangle is isosceles. [2 marks]

8 Two points A and B have position vectors:
   $\mathbf{a} = 2\mathbf{i} + \mathbf{j}$
   $\mathbf{b} = -3\mathbf{i} + 5\mathbf{j}$.
   The point C lies
   • above the x axis
   • on the line $x = 2$.
   The length of the vector $\overrightarrow{AC}$ is half the length of the vector $\overrightarrow{BC}$.
   Find the position vector of C. [6 marks]

9 Vectors $\mathbf{a}$, $\mathbf{b}$ and $\mathbf{c}$ are defined by:
   $$\mathbf{a} = \begin{pmatrix} 1 \\ 3 \end{pmatrix}$$
   $$\mathbf{b} = \begin{pmatrix} 4 \\ -2 \end{pmatrix}$$
   $$\mathbf{c} = \begin{pmatrix} 2 \\ 5 \end{pmatrix}$$
   (i)   Draw a diagram to show $\mathbf{a} + \mathbf{b}$. [1 mark]
   (ii) Work out $\mathbf{b} - \mathbf{a} + \mathbf{c}$ in column vector form. [1 mark]
   For some constants m and n, $\mathbf{c} = m\mathbf{a} + n\mathbf{b}$.
   (iii) Work out the exact values of m and n. [2 marks]

10 Vector **a** has magnitude $\sqrt{32}$ and bearing 135°.

    (i)   Find **a** in column vector form.   [2 marks]

    Vector **b** has the x-component -2 and the y-component 4.

    (ii) Find the magnitude and direction of **b**, giving the direction as an angle between $-180°$ and $180°$, measured anticlockwise from the positive x axis.   [3 marks]

    Point A has position vector **a**, and point B has position vector **b**.

    The position vector of point C has the x-component 3.

    (iii) If $\overrightarrow{AC}$ is parallel to **b**, find the coordinates of C.   [2 marks]

    (iv) If, instead, A, B and C are collinear, find the coordinates of C. [2 marks]

13 Exponentials and logarithms

1   Solve the equation $3^x = \frac{1}{81}$.   [2 marks]

2   Solve the equation $8^x = \frac{1}{4}$.   [2 marks]

3   Solve the equation $\log_2 x = 5$.   [2 marks]

4   Simplify $\log_a a$.   [1 mark]

5   Simplify $\log_a \sqrt{a}$.   [1 mark]

6   Simplify $\log_a \frac{1}{a}$.   [1 mark]

7   Write down the inverse function of $y = a^x$.   [1 mark]

8   Solve the equation $3^x = 10$.   [2 marks]

9   Sketch the graph of $y = e^x$.   [1 mark]

10 Sketch the graph of $y = e^{-x}$.   [1 mark]

11 Solve the equation $1.05^x = 20$.   [2 marks]

12 n is an integer.
    What is the greatest value of n such that $1.2^n < 10$?   [3 marks]

13 Write down the derivative of e^{4t}.   [1 mark]

14 Express $\log_a xy - \log_a x^2$ as a single logarithm.   [2 marks]

15 An antique diamond necklace was valued in January 2001 at £1200.
    It has been increasing in value by 4% each year.

    (i)   How much will it be worth 20 years later, in 2021?
              Give your answer to the nearest pound.   [2 marks]

    (ii) Write down an expression for its value, V, t years after its
              valuation.   [1 mark]

    (iii) In which year was the necklace worth £1000. Assume that the
              rate of increase has been constant for a long time.   [3 marks]

16 Make x the subject of the following formulae.

    (i) $5e^{3x} + a = b$ [3 marks]

    (ii) $\ln x - \ln(x + 1) = \ln y$ [3 marks]

17 A curve has the equation $y = 7 - 8e^{-x}$.

    (i)   Sketch the graph of the curve, showing the coordinates of the
              points where the graph crosses the y axis and the behaviour of
              the graph for large values of x.   [3 marks]

    (ii) A point on the curve has the coordinates $(x, 6.5)$.
              Find the exact value of x.   [3 marks]

18 A Christmas tree farmer models the growth of his trees in two different
    ways. He buys the trees as 3-year-old seedlings, so he does not need his
    model to be valid for the first 3 years.

    In model A, the farmer uses $h = -1.9 + 1.6 \ln t$, where h is the height in
    metres and t is the age of the tree in years.

(i) Calculate the predicted height of a tree at 5 years and at 55 years, giving your answers to the nearest centimetre. [2 marks]

(ii) Explain why the model may not be valid as t gets very large. [1 mark]

(iii) Explain why the model is not valid for $t = 0$. [1 mark]

In model B, the farmer uses $h = 4.6 - 6.5e^{-0.1t}$.

(iv) Calculate the predicted height of a tree at 5 years and at 55 years, and compare these answers with your answers from part (i). [3 marks]

(v) What does the model predict for the height of the tree for large values of t. [1 mark]

(vi) Explain why this model is not valid for $t = 0$. [1 mark]

(vii) Find the gradient function for model B, and evaluate it when $t = 5$. Explain the significance of your answer. [4 marks]

From experience, the farmer knows that a 20-year-old tree is about 3 m tall.

(viii) Determine which of the two models is the better model for this situation. [2 marks]

19 Raoul ties a small heavy object to a string.
He adjusts the length of the string and counts how many times it will swing in 30 seconds, counting only completed swings.
The table shows his results.

Length (x m)	0.3	0.5	0.8	1.0	1.2	1.5
Number of swings in 30 seconds	27	21	16	14	13	12

Raoul thinks the number can be modelled by $N = kx^n$, where N is the number of completed swings, x is the length of the string in metres, and k and n are constants.

(i) Show how plotting $\ln N$ against $\ln x$ can be used to test the model. [3 marks]

(ii) Complete the table, giving the values to 2 decimal places. Plot the graph. [3 marks]

$\ln x$						
$\ln N$						

(iii) Use the graph to find the constants k and n. [3 marks]

(iv) Calculate how many swings the model predicts for a string of length 0.75 m. Give your answer to 1 decimal place. [2 marks]

(v) Find the length of the string so that exactly 15 completed swings occur in 30 seconds. [2 marks]

20 The table shows the population of the UK in the nineteenth century from census data.

Year	1851	1861	1871	1881	1891
Population (millions)	27.368	28.918	31.435	34.935	37.802

The population can be modelled by the equation $P = Ae^{kx}$, where P is the population in millions and x is the number of years after 1800.

(i) Show that the model can be written in the form
$\ln P = \ln A + kx$. [2 marks]

(ii) Explain how a graph of $\ln P$ against x can test whether the model is valid. [1 mark]

(iii) Complete the table showing the values of x and $\ln P$. [2 marks]

x	51				91
$\ln P$	3.3094				

(iv) Plot a graph of $\ln P$ against x.
Determine whether the model is valid. [2 marks]

(v) Use the graph to find the constants A and k.
Explain the significance of A in the context of the question. [3 marks]

(vi) Use your model to predict the population in 1901 and in 1951. [2 marks]

(vii) The actual values of the population were 38.237 million in 1901 and 50.225 million in 1961.
Comment on the suitability of the model for the population in the twentieth century. [2 marks]

(viii) Find the gradient function and its value when $t = 91$.
Interpret your answer in the context of the question. [4 marks]

14 Data collection

Questions in Chapter 14 refer to the large data set, CIA World Factbook, used by MEI for its specimen and practice papers.
This is different from the data set(s) used for live papers.

It can be found at:

www.ocr.org.uk/qualifications/as-a-level-gce-mathematics-b-mei-h630-h640-from-2017

1 Give one advantage and one disadvantage of systematic sampling. [2 marks]

2 What is a simple random sample? [1 mark]

3 Give one advantage and one disadvantage of quota sampling. [2 marks]

4 What is a disadvantage of doing a census over taking a sample? [1 mark]

5 Kerry is doing a survey of prices for 'coffee to go'.
   She uses a sample from shops in her local high street.
   What is the population for her survey? [1 mark]

6 Jon selects a sample of 40 countries from the CIA World Factbook.
   He plans to investigate health expenditure as a percentage of GDP.

   (i)   Using your knowledge of the large data set, give two reasons why he may find it difficult to obtain a representative sample. [2 marks]

   (ii) Suggest a strategy for dealing with each of the difficulties identified in (i). [2 marks]

   (iii) Explain how Jon could collect a stratified sample of 40 countries from the large data set. [2 marks]

7 Shani plans to take a random sample of 30 countries from the large data set, CIA World Factbook.

   (i)   Describe how she could select her sample. [2 marks]

   (ii) Describe a problem she may encounter, and briefly explain how to deal with it. [1 mark]

8 Xavier says 'I shall take a cluster sample from the large data set by selecting one country from each region.'

   (i)   Explain why this is not a cluster sample. [2 marks]

   (ii) Name the sampling method he describes. [1 mark]

9 Ashraf is investigating the ways people do their shopping.
   He works in a corner shop on Saturdays and plans to ask some of the customers to fill out a short questionnaire about their shopping habits.
   Critically analyse his strategy. [3 marks]

10 The table shows the ages of 255 members of a tennis club.

Age	Frequency
14 to 18	39
19 to 30	90
31 to 65	126

Work out how many of each age group should be included in a stratified sample of 50. [3 marks]

15 Data processing, presentation and interpretation

1   Diane is looking at data about weather in her local area for the previous 4 years.
    She finds the rainfall was sometimes recorded as 'tr' and other times 'n/a'.
    Given that she wants to calculate averages, advise her how to deal with
    these items.   [2 marks]

2   For which of the following would it make sense to calculate the mean?
    height   colour   shoe size   passport number   [2 marks]

3   In which of these diagrams is the mode less than the mean?   [1 mark]

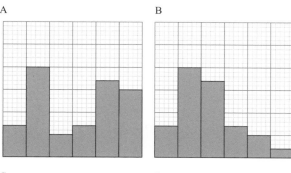

4   In which of these diagrams is the median less than the mean?   [1 mark]

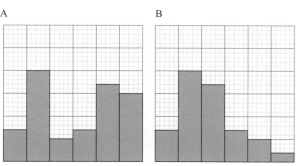

C   D

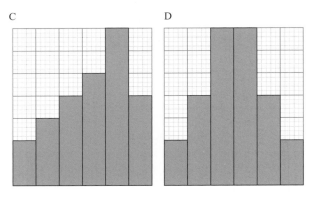

5   Calculate the mean, mode and median of the following data.
    2, 5, 4, 6, 5, 2, 2, 5, 6, 8, 1, 2   [2 marks]

6   Describe the correlation shown by these data, explaining your
    answer.   [2 marks]

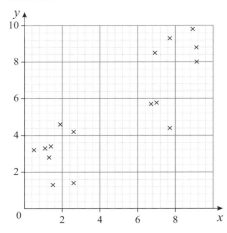

[2 marks]

7   After looking at the pie chart, Mundher says 'There are 90 more people
    who prefer to play tennis than rounders.'

    There are 100 people represented by the whole pie chart.
    Explain Mundher's error.   [2 marks]

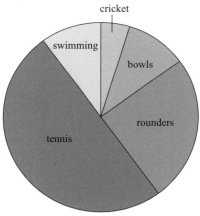

8 The graph shows the scores of a group of 20 people on a reaction time test.

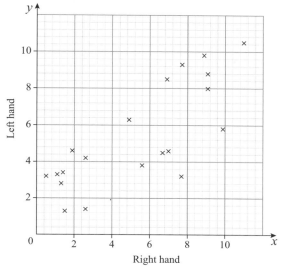

(i) Describe the correlation in the scatter diagram. [1 mark]

(ii) Jan has a score of 4 using her right hand.
Estimate her likely score using her left hand.
Comment on the reliability of your estimate. [2 marks]

9 Calculate the standard deviation of the data with the following summary data.

$$\sum x^2 = 10\,789, \sum x = 381, n = 15$$ [3 marks]

10 Calculate the standard deviation of the following data, and identify any outliers.

12, 15, 6, 16, 11, 15, 12, 14, 13, 11 [4 marks]

11 The following graph represents scores on a national test.

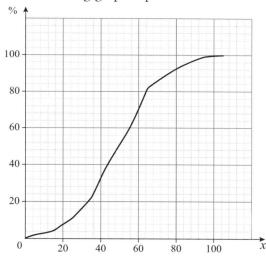

(i) The pass mark is set at the 70th percentile.
What is the pass mark? [1 mark]

(ii) Write down the range of marks obtained by the middle 80% of the candidates. [2 marks]

12 The histogram and table show the adult masses of 30 snow leopards.

Mass (kg)	Frequency
	4
$40 \leqslant m < 50$	14
$60 \leqslant m < 75$	

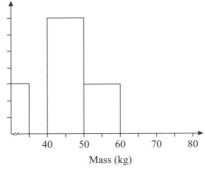

Estimate the number of the snow leopards that weigh more than 55 kg. [6 marks]

13 The population of polar bears is threatened by the reduction in sea ice.

The box and whisker plot shows the results of a survey taken 10 years ago of the masses of adult polar bears.

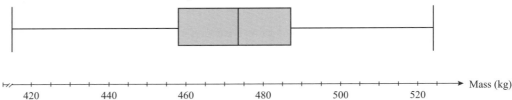

A more recent survey of the masses of ten polar bears gave the following results.

461   474   458   450   446   457   484   484   455   465

(i) Calculate the median and quartiles of the data in the more recent survey. [2 marks]

(ii) Use the results of your calculations in part **(i)** to compare the two sets of data. [4 marks]

(iii) Is there evidence to suggest that polar bears are getting thinner? You should justify your answer. [2 marks]

(iv) Are there any outliers in the data from the earlier survey? Show calculations to justify your answer. [2 marks]

14 A   B   C

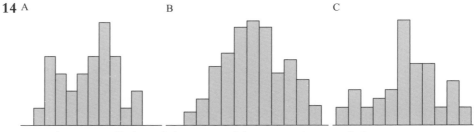

Two dice are rolled, and the sum of the scores is recorded.

The graphs show the results for 30 throws, 50 throws and 100 throws.

Match the graphs to the numbers of throws, explaining your choices. [2 marks]

15 Noah is writing a report about the weather in the local area for the past 3 years. He focuses on the rainfall and is trying to decide which of the following two graphs to use to illustrate his report.

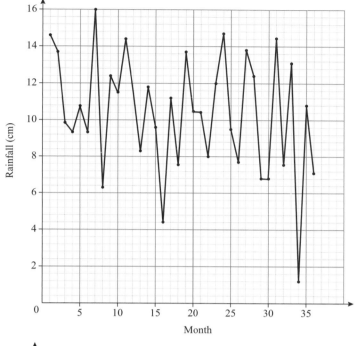

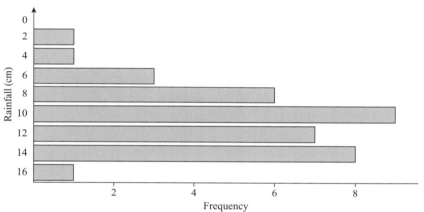

Advise Noah on which graph would show the data better. [3 marks]

16 The cumulative frequency graphs show the results from Paper 1 and Paper 2 of a GCSE Mathematics paper.

The solid line shows the marks from Paper 1, and the dashed line shows the marks from Paper 2.

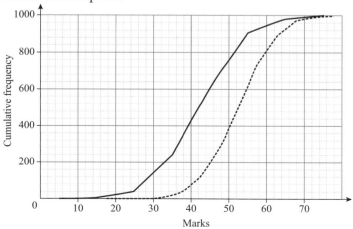

Which paper was harder?

Justify your answer by referring to measures estimated from the graphs. [4 marks]

17 The scatter graph shows information about the height at 16 years old and monthly salary at 25 years old of a group of 20 people.

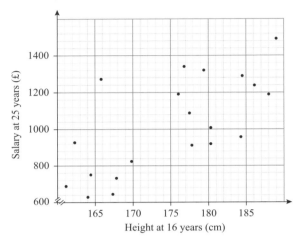

(i) Describe the correlation in the scatter diagram. [1 mark]

(ii) Estimate the salary at 25 years old of someone who is 172 cm tall at the age of 16. [1 mark]

(iii) Estimate the salary at 25 years old of someone who is 140 cm tall at the age of 16. [1 mark]

(iv) Which of your estimates in parts (ii) and (iii) is likely to be the better estimate? Justify your answer. [2 marks]

(v) Do you think a person's height affects how much they are paid? Justify your answer. [2 marks]

18 Compare the life expectancy for citizens of the UK with those of their three nearest neighbours on the continent of Europe.

Country	Population	Life expectancy
Netherlands	17 million	81.1
Belgium	10 million	79.9
France	66 million	81.7
UK	64 million	80.4

You should calculate measures to support your conclusions. [4 marks]

19 Mary can buy eggs from two different suppliers for her business making cupcakes.

Samples from the two suppliers yield the following information:

• Supplier A has eggs with a mean mass of 61 g and a standard deviation of 10.2 g.

• Supplier B has ten eggs with these masses (in grams): 53, 57, 68, 56, 60, 48, 70, 40, 56, 53.

Advise Mary on which supplier to use.

Use the results of any calculations you make to support your recommendation. [4 marks]

20 The diagram shows the heights of bushes of two different species of hydrangea.

Species A		Species B
7	5	
9 9 8	6	0
9 9 7 6 5 5 2 0 0 0	7	1 6 9 9
6 5 5 5 5 4 4 4 3 3 1 0	8	1 1 1 1 2 3 3 5 5 6 7 8 8 9
8 2 0 0	9	2 3 4 5 7 7 8 8
	10	4 6 7

Key $10 \mid 4 = 104$ cm

(i) Compare the heights of bushes of species A and B, with reference to appropriate measures of central tendency and spread. [5 marks]

(ii) Determine whether any of the data are outliers. [2 marks]

16 Probability

1 Find the probability of getting a square number when you throw a normal dice. [1 mark]

2 Find the probability of getting a double when you throw two normal dice. [1 mark]

3 **(i)** Find the probability of getting a total of 12 when you throw two normal dice. [1 mark]

   (ii) How many totals of 12 would you expect to get in 180 throws? [1 mark]

4 The probability of getting various outcomes when a spinner is turned is as shown:

Outcome	Probability
Red	0.2
Green	0.1
Yellow	0.4
Blue	0.1

   The rest of the time you get black.

   Find the probability that you do not get yellow. [1 mark]

5 How many different groups of two people can be selected from a group of five people? [2 marks]

6 Find the probability of getting at least one head when six coins are thrown. [2 marks]

7 A dice is thrown three times.

   Find the probability of getting exactly one 5. [2 marks]

8 A bag contains three pink, four green and six blue balls.
   The balls are identical except for their colour.

   Two balls are removed from the bag without replacement.
   Determine the most likely outcome. [3 marks]

9 A bag contains three red balls, four green balls and five yellow balls.

   If two balls are taken from the bag together (without replacement),
   find the probabilities of the following scenarios.

   (i) The balls are different colours. [3 marks]

   (ii) At least one of the balls is yellow. [2 marks]

10 The integers 1 to 10 are placed in one or more of the following categories:

   E: even numbers

   O: odd numbers

   P: prime numbers

   S: square numbers

   One of the numbers is chosen at random. Find the probability of
   it falling within the following sets.

   (i) P' [2 marks]

   (ii) $P \cup E$ [2 marks]

   (iii) $O \cap S$ [2 marks]

11 A school council of five members is to be chosen by drawing names out of a hat. There are four boys and six girls to choose from.

Simplifying your answers, find the probabilities of the following scenarios.

  (i)   All the council members are girls.   [3 marks]

  (ii) Exactly two of the council members are boys.   [3 marks]

12 (i) Two fair dice are thrown together, and their scores are denoted by X and Y.

  By creating a sample space diagram, determine the probability distribution of $|X - Y|$.   [4 marks]

  (ii) Find the expected number of 1s in 180 throws.   [2 marks]

13 The results of a survey are shown in the table below. The figures are percentages.

	Likes popcorn	Doesn't like popcorn
Goes to the cinema	12	27
Doesn't go to the cinema	43	18

Show that the events 'Likes popcorn' and 'Goes to the cinema' are not independent.   [4 marks]

17 The binomial distribution

1 Jo is practising taking penalty kicks.
On average, she scores 13 times out of 20 attempts.
X is the number of successful shots.
State the distribution of X, including the value of any parameters. [2 marks]

2 Jim throws a normal dice five times.
Find the probability that he throws no 3s. [2 marks]

3 Sean sends out 60 invitations to an event by using social media.
He expects a 35% response rate.
X is the number of responses he gets.
State the distribution of X, including the value of any parameters. [2 marks]

4 Annie shoots at a target.
Y is the number of shots that hit the target.
$Y \sim B(50, 0.6)$.
Find her mean number of hits. [1 mark]

5 Jed throws a pair of dice 100 times.
Y is the number of double 1s he gets.
State the distribution of Y, including the value of any parameters. [2 marks]

6 When Sue plays tennis, she gets her first serve in 60% of the time.
In one set, she serves 20 times.
Find the probability that she gets no more than one first serve in. [3 marks]

7 Xena is running a campaign and sends out 50 emails to businesses in her town.
Past experience shows that 15% of businesses reply.
X is the number of businesses that reply to this campaign.

   (i) State the distribution of X, giving the value of the parameters. [2 marks]

   (ii) Calculate the probability that at least ten businesses reply. [1 mark]

   (iii) Find the expected number of replies. [1 mark]

   (iv) Find the value of X that is most likely, showing clearly how you obtained your answer. [3 marks]

8 Amy and Bayan play a game.
They throw five bean bags each and score a point for every one that lands in a hoop.
The probability that either child gets the bean bag in the hoop is $\frac{1}{3}$, whether the previous throw is a success or not.

   (i) Explain why the score can be modelled using a binomial distribution. [2 marks]

   (ii) Complete the table of the probabilities for each score. [2 marks]

Score	0	1	2	3	4	5
Probability	0.1317	0.3292				0.0041

   (iii) Find the probability that at the end of the game the scores are tied. [2 marks]

   (iv) Calculate the probability that Amy wins. [2 marks]

9 Ahmed thinks that he gets stuck in a traffic jam on average once a week on his way to work.

He models the number of days he gets stuck per week using a binomial distribution with $n = 5$ and $p = 0.2$.

(i) Find the probability that he goes a week without a traffic jam. [2 marks]

(ii) Calculate the probability that he is stuck once in the first week and once in the second. [3 marks]

(iii) Calculate the probability that he is stuck twice in two weeks. [3 marks]

(iv) Explain why the answers to parts (ii) and (iii) are not equal. [1 mark]

10 The manager of a supermarket checks boxes of eggs.

She believes that the number of broken eggs in a box fits a binomial distribution with $n = 6$ and probability p.

She usually finds that 90% of boxes of six eggs have no broken eggs.

(i) Use trial and improvement to calculate the probability that any one egg is broken, giving your answer to 4 decimal places. [3 marks]

One day she notices a box with four broken eggs.

(ii) Use the binomial distribution to calculate the probability that this happens.
Comment on your answer. [2 marks]

She suspects that the number of broken eggs does not fit a binomial distribution.

(iii) Which of the modelling assumptions may not be appropriate in this context? [2 marks]

11 The probability distribution of a discrete random variable X is given by $P(X = r) = kr^2$ for $r = 1, 2, 3$ and 4.

(i) Determine the value of k. [2 marks]

Two values of X are chosen at random.

(ii) Find the probability that they are equal. [3 marks]

18 Statistical hypothesis testing using the binomial distribution

1 It is suspected that more boys than girls are born immediately after a war. A hypothesis test is carried out.

Write down the null hypothesis. [1 mark]

2 It is suspected that a six-sided dice is biased in favour of showing a 6. A hypothesis test is carried out.

Write down the null and alternative hypotheses. [2 marks]

3 In a hypothesis test, the null hypothesis is

$H_0 : p = 0.7$.

Write down the alternative hypothesis if the test is two-tailed. [1 mark]

4 A hypothesis test has

$H_0 : p = 0.6$

$H_1 : p > 0.6$.

Find the critical region if the test is carried out at the 5% significance level with 15 trials. [3 marks]

5 The significance level of a one-tailed hypothesis test is 5%. The test statistic is calculated as 0.035.

What is the outcome of the test? [2 marks]

6 A hypothesis test has

$H_0 : p = 0.45$

$H_1 : p \neq 0.45$.

Find the critical region if the test is carried out at the 5% significance level with ten trials. [3 marks]

7 A butcher surveys his customers who buy sausages on a Saturday, and 20% of those customers say they are having a barbeque.

He decides to carry out a hypothesis test to find out whether customers are less likely to have a barbeque on a rainy Saturday.

He uses a 10% significance level.

(i) Write down the null and the alternative hypotheses that the butcher should use.

Give a reason for your alternative hypothesis. [3 marks]

He surveys 20 customers buying sausages on a rainy Saturday and finds that only one customer is having a barbeque.

(ii) Conduct the hypothesis test, stating the conclusions clearly. [4 marks]

8 An amateur choir of 24 members has been tracking how many members cough during a rehearsal.

They find over a long time period that the probability of coughing is 30%.

Friedrich asks the choir members each to eat a cough sweet before a rehearsal.

He performs a hypothesis test at the 5% significance level to see whether that makes any difference to the probability of coughing during the rehearsal.

(i) Write down the null and alternative hypotheses that Friedrich should use. Explain the form of the alternative hypothesis. [3 marks]

(ii) Determine the critical region for his test. [4 marks]

(iii) He finds that only three people have coughed that rehearsal. Complete the test. [2 marks]

9 Winston hopes to be voted school captain by his class.
He thinks each of the children in his class of 30 is equally likely to vote for him or his rival.

(i) Explain the modelling assumptions needed for the number of children who vote for Winston to fit a binomial distribution. [2 marks]

He buys sweets for his classmates to try to win their support.
A total of 20 children vote for Winston.

(ii) Carry out a test at the 5% significance level to see whether there is any evidence that buying sweets has had any effect on how the children voted.
State the null and alternative hypotheses and any conclusions clearly. [8 marks]

10 Members of a residents' group have been campaigning for traffic calming measures in their village and have been collecting data.
They record that 8% of vehicles go through the village much too quickly in their opinion.

As a result of their campaign, speed bumps are installed in the village.
The residents' group performs a hypothesis test at the 10% significance level to see whether the speed bumps have improved the situation.

(i) Write down the null and the alternative hypotheses that they should use. [2 marks]

The group surveys 50 vehicles.

(ii) Determine the critical region for their test. [2 marks]

One vehicle was judged to be going much too fast.

(iii) State clearly the conclusion that the group should reach. [3 marks]

19 Kinematics

You should assume $g = 9.8\,\mathrm{m\,s^{-2}}$, unless stated otherwise.

1 What does the gradient of a distance travelled–time graph represent? [1 mark]

2 What does the gradient of a speed–time graph represent? [1 mark]

3 What does the area between the graph and the time axis of a speed–time graph represent? [1 mark]

4 Explain the difference between distance travelled and displacement. [1 mark]

5 Sketch the speed–time graph for the distance travelled–time graph in the diagram. [2 marks]

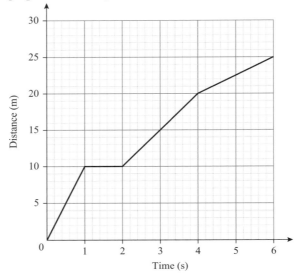

6 Sketch the distance travelled–time graph for the speed–time graph in the diagram. [2 marks]

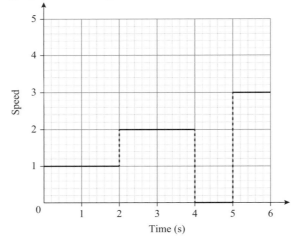

7 A particle travels 220 m in 5 seconds.
Find its average speed. [2 marks]

8 The velocity–time graph in the diagram shows the motion of a particle.

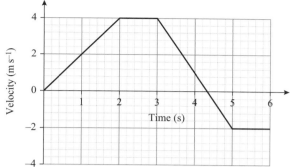

How far is the particle from its starting point at the end of 6 seconds? [3 marks]

9 Given that $a = 2\,\text{m s}^{-2}$, $s = 12\,\text{m}$ and $v = 9\,\text{m s}^{-1}$, work out u. [2 marks]

10 Given that $u = 2\,\text{m s}^{-1}$, $s = 12\,\text{m}$ and $v = 4\,\text{m s}^{-1}$, work out t. [2 marks]

11 A particle travels at $5\,\text{m s}^{-1}$ for 3 seconds and $8\,\text{m s}^{-1}$ for 5 seconds. Find its average speed. [3 marks]

12 A stone is dropped from a height of $20\,\text{m}$. How long does it take to reach the ground? [3 marks]

13 A stone is dropped and takes 5 seconds to reach the ground. From what height was it dropped? [3 marks]

14 A particle accelerates at $2.5\,\text{m s}^{-2}$ and travels $100\,\text{m}$ in 10 seconds. Find its initial velocity. [3 marks]

15 The diagram shows the displacement–time graph for a jogger as she runs along a straight track and then back to her starting position. The displacement, s, is measured in metres, and the time, t, is measured in seconds.

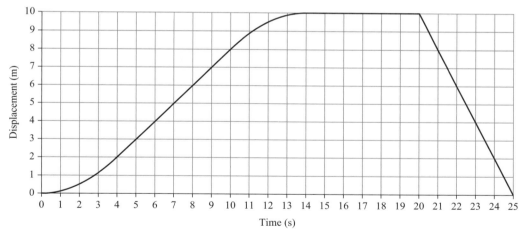

    (i) Between what times was the jogger moving with a constant velocity away from her starting position? [1 mark]

    (ii) For how long did the jogger remain stationary before she started to move back towards her starting position? [1 mark]

    (iii) What was the average speed of the jogger for the full period of time shown in the graph? [2 marks]

16 Every day Mark drives on a motorway at an average speed of $30\,\text{m}\,\text{s}^{-1}$. This morning there were delays, and Mark's average speed on the motorway was only $25\,\text{m}\,\text{s}^{-1}$.

The journey took 2 minutes longer than usual.
Work out the distance that Mark drives on the motorway. [4 marks]

17 A particle moves with constant acceleration over a period of 20 seconds. The displacement–time graph for the motion is shown below.

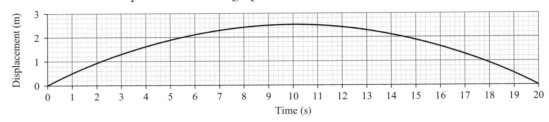

The maximum value of the displacement during the motion was 2.5 m.
Find the initial velocity and the acceleration of the particle. [5 marks]

18 A and B are two points on a straight horizontal road which are 120 m apart.
When a car passes A, its speed is $10\,\text{m}\,\text{s}^{-1}$ in the direction AB.
It then accelerates uniformly, and when it reaches B its speed is $40\,\text{m}\,\text{s}^{-1}$.

(i) Find the car's acceleration. [2 marks]

At the instant that the car passes A, a motorbike begins to accelerate from rest, at B, in the direction BA. The motorbike and the car pass each other 4 seconds later.

(ii) Find the speed of the motorbike at the time when it passes the car. [4 marks]

19 A stone is thrown upwards from the ground with an initial speed of $20\,\text{m}\,\text{s}^{-1}$. At the same moment, a second stone is dropped from rest from a position vertically above the first stone.

Both stones hit the ground at the same instant.

(i) Determine the height from which the second stone was dropped. [4 marks]

(ii) Determine the speed of the faster stone when it hits the ground. [2 marks]

20 A car accelerates from rest with an acceleration of $0.8\,\text{m}\,\text{s}^{-2}$.
After 20 seconds, another car accelerates from rest at the same position in the same direction with an acceleration of $1.8\,\text{m}\,\text{s}^{-2}$.

(i) After how long from the time that the first car starts will the cars be in the same position? [5 marks]

(ii) Determine the distance that each of the cars will have travelled at the time that they are in the same position. [2 marks]

21 The graph below is a velocity–time graph for the motion of a particle. Initially, the displacement of the particle from a point O is 10 m.

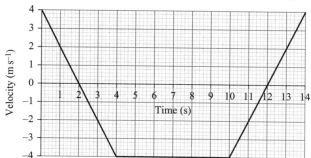

  (i) Find the acceleration of the particle during the first 4 seconds of the motion. [2 marks]

  (ii) At what time did the particle first pass through O? [4 marks]

  (iii) Determine the furthest distance away from O that the particle reaches in the 14 seconds of the motion. [4 marks]

22 Two balls are thrown into the air from ground level at the same instant. The initial speed of one of the balls is twice that of the other. Given that one of the balls hits the ground 10 seconds after the other, find the speed at which the slower ball was projected. [4 marks]

23 A monorail at a theme park transports customers from one part of the park to the other along a straight track which is 900 m long.

When trains leave, they accelerate from rest with an acceleration of $2\,\mathrm{m\,s^{-2}}$ until they reach a speed of $12\,\mathrm{m\,s^{-1}}$.

They then travel at this speed until they decelerate as they approach the end of the journey. When approaching the end of the journey, the trains decelerate at $-1\,\mathrm{m\,s^{-2}}$.

  (i) Draw a velocity–time graph to illustrate this motion. [3 marks]

  (ii) Find the total amount of time for which the train is in motion. [3 marks]

The manager of the monorail is considering increasing the constant velocity at which the trains move to $16\,\mathrm{m\,s^{-1}}$.

  (iii) How much time will be saved on the overall journey time by making this change, if the acceleration and deceleration remain at the same rate? [3 marks]

20 Forces and Newton's laws of motion

The acceleration due to gravity is denoted by $g\,\mathrm{m\,s^{-2}}$. Unless otherwise instructed, when a numerical value is needed, use $g = 9.8$.

1 State the S.I. units for mass, length and time. [1 mark]

2 A book rests on a smooth horizontal table.
Draw a diagram to show the forces acting on the book. [1 mark]

3 Forces $F_1 = 3\mathbf{i} - 4\mathbf{j}$ and $F_2 = -\mathbf{i} + 2\mathbf{j}$ act on a particle.
Find the resultant force. [2 marks]

4 A block of mass 8 kg rests on a smooth horizontal surface.
Calculate the normal reaction of the surface on the block. [2 marks]

5 A force of 2 N acts on a particle of mass 5 kg.
Calculate the acceleration of the particle due to the force. [1 mark]

6 A force acts on a particle of mass 8 kg, giving it an acceleration of $1.5\,\mathrm{m\,s^{-2}}$.
Calculate the magnitude of the force. [1 mark]

7 A force of 12 N acts on a particle, giving it an acceleration of $0.3\,\mathrm{m\,s^{-2}}$.
Work out the mass of the particle. [1 mark]

8 A mass of 3 kg is suspended in equilibrium by a string.
Draw a diagram to show the forces acting on the mass. [1 mark]

9 A force acts on a particle of mass 4 kg, giving it an acceleration of
$-0.4\mathbf{i} + \mathbf{j}\,\mathrm{m\,s^{-2}}$.
Determine the force. [1 mark]

10 A force of $6\mathbf{i} - 8\mathbf{j}\,\mathrm{N}$ acts on a particle of mass 2 kg.
Work out the magnitude of the acceleration. [2 marks]

11 Two particles, of different masses, are joined by a light inextensible string which passes over a smooth fixed pulley.
Draw a diagram to show the forces on each particle. [2 marks]

12 Two particles, one of mass 5 kg and the other of mass 7 kg, are joined by a light inextensible string which passes over a smooth fixed pulley.
Write down equations of motion for each particle. [2 marks]

13 Two particles, one of mass 4 kg and the other of mass 3 kg, are joined by a light inextensible string which passes over a smooth fixed pulley.
Find the acceleration of the particles and the tension in the string in terms of g. [3 marks]

14 A train, with driving force 1500 N and mass 2 tonnes, pulls a single coach of mass 1.5 tonnes.
The resistance to motion of the train is 600 N, and the resistance to motion for the coach is 400 N.
Determine the acceleration of the train and the force in the coupling. [4 marks]

15 A book of mass m kg rests on a table.
Use appropriate laws of motion to demonstrate that the reaction force of the book on the table is mg N. [2 marks]

16 A car is pulling a trailer using a towbar.

By means of a suitable equation, establish what conditions or assumptions must hold in order that the towbar should experience equal and opposite forces from the car and the trailer. [4 marks]

17 The following forces (in Newtons) act on a ball of mass 2 kg: $F_1 = 2\mathbf{i} + 3\mathbf{j}$ and $F_2 = -3\mathbf{i} - 4\mathbf{j}$ ($\mathbf{i}$ is horizontal, and $\mathbf{j}$ is vertically upwards).

Find the additional force that must be applied, in order that the ball falls vertically with an acceleration of $10\,\mathrm{m\,s^{-2}}$. [4 marks]

18 A team of huskies is pulling a sledge of mass 300 kg (including its cargo). A second sledge of mass 400 kg (including its cargo) is pulled along by the first sledge, by means of a light inextensible rope.

The first sledge experiences total resistance forces of 500 N, while the second sledge experiences total resistance forces of 400 N.

(i) Determine the tension in the rope when the sledges are accelerating at $0.5\,\mathrm{m\,s^{-2}}$. [3 marks]

(ii) Work out the maximum magnitude of the deceleration that is possible without the two sledges colliding. [4 marks]

19

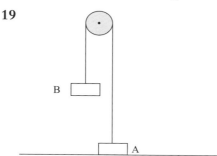

Two blocks, A and B, of masses $2m$ kg and $3m$ kg, respectively, are attached to the ends of a light inextensible rope, which passes over a smooth fixed pulley, as shown in the diagram.

A is held on the floor, and B hangs in equilibrium, at a height of 2 m above the floor.

A is released. Find the time taken for B to hit the floor. [6 marks]

20

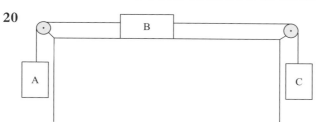

Initially, the block B of mass 3 kg is held at rest on a rough table. At one end, it is connected by a light inextensible rope, via a smooth pulley, to the block A of mass 2 kg, which hangs down one side of the table.

At its other end, B is connected by another light inextensible rope, via a smooth pulley, to the block C of mass 5 kg, which hangs down the other side of the table. (See diagram.)

Block B is then released.

If the tension in the rope connecting B and C is twice the tension in the rope connecting B and A, determine, in terms of g:

(i) the acceleration of the blocks [6 marks]

(ii) the frictional force between B and the table. [2 marks]

21 A lift of mass 500 kg contains a man with a mass of 70 kg.

The man has a parcel which he holds by a vertical string (assumed to be inextensible and have negligible mass), so that the parcel is suspended above the floor of the lift. (The mass of the lift does not include the mass of the man or the parcel.)

The lift is suspended vertically by a cable.

The tension in the cable is 5300 N, and the tension in the string is 9 N. Find:

  (i) the acceleration of the lift   [6 marks]

  (ii) the mass of the parcel.   [3 marks]

22 Some military equipment is delivered by parachute, as illustrated in the diagram.

The parachute is of mass 10 kg and is subject to air resistance of 600 N.

Box A is of mass 50 kg and is subject to air resistance of 300 N.

Box B is subject to air resistance of 400 N. The ropes between the boxes and the parachute are light and inextensible.

Given that the parachute and boxes are moving at constant speed, find the tensions in the two ropes and the mass of box B.   [9 marks]

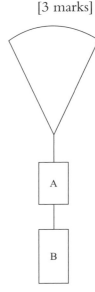

23 A lift of mass 400 kg is suspended by a cable that can bear a tension of up to 10 000 N.

If the lift is to accelerate at up to 2 m s^{-2}, what is the maximum number of passengers that it can carry?

Assume that all passengers have a mass of 80 kg.   [8 marks]

21 Variable acceleration

1 The displacement, in metres, of a particle at time t seconds is given by $s = 5t^3 - t^2$. Find an expression for its velocity in terms of t. [1 mark]

2 The velocity of a particle at time t seconds is given by $v = 3 - 2t^2$. Find an expression for its acceleration in terms of t. [1 mark]

3 A particle starts from rest, and its acceleration at time t seconds is given by $a = \frac{1}{2}t - 5$.
Find an expression for its velocity at time t seconds. [2 marks]

4 A particle leaves point A, and its velocity t seconds later is given by $v = 0.1t^2 - 2t$.
Find an expression for its distance from A at time t seconds. [2 marks]

5 A particle passes point A with a velocity of $3\,\mathrm{m\,s^{-1}}$.
Its acceleration at time t seconds is given by $a = 4 - t$.
Find an expression for its velocity at time t seconds. [2 marks]

6 A particle moves so that its distance from A at time t seconds is given by $s = 4t^3 - t^2 + 1$.
How far is it from A after 5 seconds? [1 mark]

7 A particle starts from rest 3 m from B and travels with acceleration given by $a = 3t - 5$.
Determine an expression for its distance from B at time t seconds. [3 marks]

8 Show that $s = ut + \frac{1}{2}at^2$ can be obtained, using calculus, from the equation: acceleration $= a$, $s =$ the displacement after t seconds, and $u =$ the initial velocity. [4 marks]

9 The displacement of a particle at time t seconds is given by $s = t^3 - 30t^2 + 10t$.

　(i) Find the velocity of the particle after 10 seconds. [3 marks]

　(ii) After how many seconds does the particle have the same velocity as its original velocity? [3 marks]

10 A particle moves in such a way that its velocity at time t seconds is given by $v = 3t^2 - 12t + 9$ for $0 \leqslant t \leqslant 3$.

　(i) Show that the particle has returned to its starting position at the end of the 3 seconds. [3 marks]

　(ii) Find the total distance travelled by the particle in the time until it returns to its starting position for the first time. [4 marks]

11 The displacement of a particle against time is shown by the following graph.

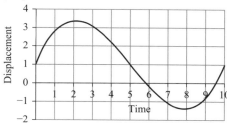

Sketch a graph of the velocity of the particle against time. [2 marks]

12 The acceleration of a particle at time $t \geqslant 0$ is given by $a = 12t - 4$.

At $t = 0$, the displacement of the particle from a fixed point O is 8 m, and it is moving with a speed of $8\,\text{m}\,\text{s}^{-1}$ towards O.

(i) Find an expression for the displacement of the particle from O at time t seconds. [5 marks]

(ii) Find the values of t when the particle is at O. [2 marks]

Answers

1 Proof

1 $n + 1$, $n + 2$

2 $2n + 1$

3 $P \Rightarrow Q$

4 $n = 9$, $n = 10$ or $n = 13$.
   There are others.

5 $3^3 = 27$

6 This is a proof question, so a short answer cannot be given. Go online to see the worked solution and mark scheme.

7 **(i)** This is a proof question, so a short answer cannot be given. Go online to see the worked solution and mark scheme.

   (ii) When the first number is even.

8 **(i)** This is a proof question, so a short answer cannot be given. Go online to see the worked solution and mark scheme.

   (ii) $c = -\dfrac{m^2}{4}$

9 **(i)** $a = \dfrac{2b}{b - 2}$

   (ii) For example, $b = 1$ gives a negative value for a, and $b = 2$ results in undefined value for a.

10 $ae \neq bd$

11 This is a proof question, so short answers cannot be given. Go online to see the worked solutions and mark scheme.

2 Surds and indices

1 $5\sqrt{3}$

2 $\sqrt{2}$

3 $\dfrac{5}{7}\sqrt{7}$

4 $\dfrac{5}{3}$

5 $\dfrac{8}{27}$

6 $5\sqrt{14}$

7 $3 + 2\sqrt{2}$

8 **(i)** 2

   (ii) 1

   (iii) $\dfrac{49}{25}$

9 $\dfrac{b^2}{2}$

10 a

11 This is a show-that question, so a short answer cannot be given. Go online to see the worked solution and mark scheme.

12 (i) This is a show-that question, so a short answer cannot be given. Go online to see the worked solution and mark scheme.

(ii) $x = \dfrac{-1 + \sqrt{5}}{2}$

(iii) This is a show-that question, so a short answer cannot be given. Go online to see the worked solution and mark scheme.

3 Quadratic functions

1 $\dfrac{1 \pm \sqrt{5}}{2}$

2 $(2x + y)(2x - y)$

3 $1.215, -0.549$

4 $(x + 3)^2 - 14$

5 $\pm \dfrac{3}{2}$

6 $(1, 5)$

7 ± 2

8 $x = -3$

9 (i) $\left(\dfrac{1}{4}, -\dfrac{25}{8} \right)$

(ii)

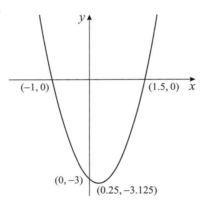

10 $k > 2\sqrt{6} - 3$ or $k < -2\sqrt{6} - 3$

11 (i) $(2x - 3)(5x + 4)$

(ii) $x = \dfrac{3}{2}$ or $x = -\dfrac{4}{5}$

(iii) $x = \pm \dfrac{\sqrt{6}}{2}$

(iv) $x < -\dfrac{4}{5}$ or $x > \dfrac{3}{2}$

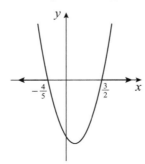

12 (i) $\quad y = 2x^2 + x - 1$

(ii)

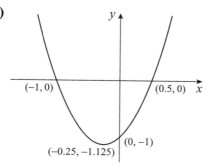

4 Equations and inequalities

1 $-1, 0, 1, 2, 3, 4$

2 $-6 \leqslant x < -2$

3 $x = 2.5, y = -2$

4 The unshaded part is the required region, as labelled.

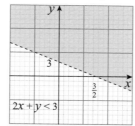

5 $x = 3, y = -1$ and $x = \dfrac{13}{5}, y = -\dfrac{9}{5}$

6 $x > 3, x < -3$

7 $-2 \leqslant x \leqslant 3$

8 $\left(\dfrac{2}{25}, \dfrac{43}{25}\right)$

9 $x < \dfrac{1}{3}$

10 $M + C = 100, 3M + 7C = 525$

$M = 43.75$ miles and $C = 56.25$ miles

11 $(1, -3)$ and $\left(\dfrac{45}{11}, \dfrac{1}{11}\right)$

12 $\{a : a > 3\} \cup \{a : a < -1\}$

5 Coordinate geometry

1 $-\dfrac{2}{3}$

2 $(-2, 4.5)$

3 13

4 -3

5 $y = 2x + 2$

6 Centre $(-4, 1)$ and radius $\sqrt{7}$

7 $(5, 0)$ and $(0, 2)$

8 This is a show-that question, so a short answer cannot be given. Go online to see the worked solution and mark scheme.

9 This is a show-that question, so a short answer cannot be given. Go online to see the worked solution and mark scheme.

10 $\left(-\frac{2}{9}, \frac{31}{9}\right)$

11 2

12 $x = \pm\frac{\sqrt{7}}{2},\ y = \frac{13}{4}$

13 This is a show-that question, so a short answer cannot be given. Go online to see the worked solution and mark scheme.

14 $3y + 2x + 13 = 0$

15 (i) $2x - y - 4 = 0$

 (ii) This is a show-that question, so a short answer cannot be given. Go online to see the worked solution and mark scheme.

 (iii) $3\sqrt{5}$

 (iv) 7.5

 (v) $\sqrt{5}$

16 (i) 5

 (ii) $(x - 5)^2 + (y - 3)^2 = 25$

 (iii) This is a show-that question, so a short answer cannot be given. Go online to see the worked solution and mark scheme.

 (iv) $(2, 7), (8, -1)$

17 (i) $y - 7 = 2(x - 3)$

 (ii) $(0, -10), (7.5, 0)$

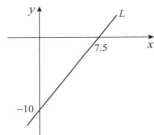

 (iii) $(-16.5, -32)$

18 (i) $x + y - 12 = 0$

 (ii) Because it must lie on the perpendicular bisector of BC, which is $x = 5$.

 (iii) $(5, 7)$

 (iv) $(x - 5)^2 + (y - 7)^2 = 52$

 (v) $k = 7 \pm \sqrt{51}$

19 (i) $(1, 3)$ or $(5, -5)$

 (ii) $(-1.4, 7.8), (5, -5)$

 (iii) $(5, -5)$

20 $\left(6\frac{4}{13}, 6\frac{6}{13}\right)$

6 Trigonometry

1 **(i)** OC

   (ii) BD

2 $30°$

3 $108°$

4 $136°$

5 $259°$

6 $36.3°$

7 $94.1°$

8 29.5 cm^2

9 $\dfrac{15\sqrt{3}}{2} \text{ cm}$

10 12.1 cm

11 $41.8°$

12 $\theta = 15°, 75°, -165° \text{ and } -105°$

13 $\theta = 60°, 660° \text{ and } 780°$

14 $\theta = 30°, 60°, 120°, 150°, 210°, 240°, 300° \text{ and } 330°$

15 $98.2 \text{ cm}^2 \text{ or } 61.4 \text{ cm}^2$

16 $144°$

17 **(i)** $-\cos\theta + \cos\theta = 0$

   (ii) $\cos\theta + \sin\theta - \cos\theta = \sin\theta$

   (iii) $-\sin\theta + \sin\theta - \cos\theta - \cos\theta = -2\cos\theta$

18 This is a show-that question, so a short answer cannot be given. Go online to see the worked solution and mark scheme.

7 Polynomials

1 $3x^3 - x^2 - 5x - 3$

2 $2x^3 - 11x^2 + 8x - 15$

3 $m + n$

4 $m - n$

5 0 is the value of p that cannot occur.

6 $3x^2 - x + 4$

7 **(i)** This is a show-that question, so a short answer cannot be given. Go online to see the worked solution and mark scheme.

   (ii) $f(x) = (x - 2)(x + 4)(2x + 3)$

   (iii)

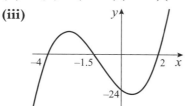

8 There is no integer root.

9 (i) $a = -4$, $b = -11$

(ii)

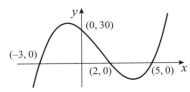

10 (i)

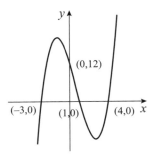

(ii) This is a show-that question, so a short answer cannot be given. Go online to see the worked solution and mark scheme.

(iii) $-4, 0, 3$

11 (i) $y = a(x + 3)(x - 1)(2x - 5)$

(ii) This is a show-that question, so a short answer cannot be given. Go online to see the worked solution and mark scheme.

12 (i)

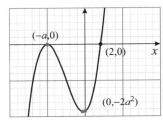

(ii) $a = 5$

8 Graphs and transformations

1 Reflection in the x axis.

2 $y = (x - 3)^2$

3 Stretch of scale factor 2 parallel to the y axis.

4 $(x + 2)^2 + (y - 5)^2 = 5$

5 $y = e^{-x}$

6 Translation of $\begin{pmatrix} 90° \\ 0 \end{pmatrix}$.

7 $y = x^3 - 4x^2 + 5x - 3$

8 $y = \dfrac{1}{x^2}$

9 $y = x^3 + x^2 - 6x$

10

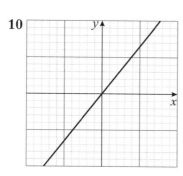

11

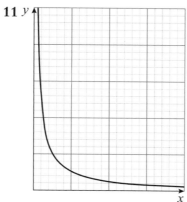

12 (i)

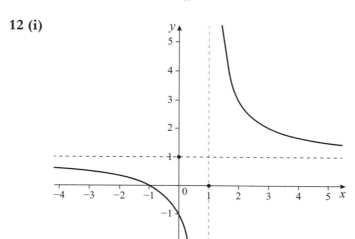

(ii) Stretch of scale factor 2 parallel to the y axis, followed by a translation of $y = \frac{1}{x}$ by $\begin{pmatrix} 1 \\ 1 \end{pmatrix}$.

13 $y = \frac{1}{10}(x+3)(x+1)(x-2)(x-5)$

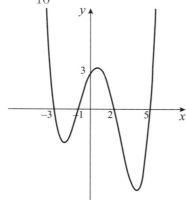

14 $y = x^3$

15 (i) Translation of $\begin{pmatrix} 0 \\ 3 \end{pmatrix}$.

(ii) Stretch of scale factor 3 parallel to the x axis.

(iii) Stretch of scale factor $\frac{1}{3}$ parallel to the y axis.

(iv) Translation of $\begin{pmatrix} -3 \\ 0 \end{pmatrix}$.

(v) Stretches of scale factor 3 parallel to the x axis and $\frac{1}{3}$ parallel to the y axis, then translation of $\begin{pmatrix} -1 \\ 3 \end{pmatrix}$.

16 $a = 4.0$ (1 d.p.)

$b = 1.5$ (1 d.p.)

17 $h = 60\sin(12t - 90°) + 75$

18

Function	Period	Coordinates of maximum closest to (0,0)	Coordinates of minimum closest to (0,0)
$y = 5\sin 2x + 1$	180°	(45°, 6)	(−45°, −4)
$y = 3\cos\left(\frac{x}{2} - 30°\right) + 5$	720°	(60°, 8)	(−300°, 2)
$y = 2\sin(4x) + 1$	90°	(22.5°, 3)	(−22.5°, −1)

9 The binomial expansion

1 30

2 10

3 729

4 18

5 84

6 1080

7 **(i)** $32 - 240x + 720x^2 - 1080x^3$

(ii) 29.671 (5 s.f.)

8 $1 - 7x + 9x^2 + 27x^3 - 54x^4$

9 56

10 4.54×10^{-6}

10 Differentiation

1 It is the gradient function.

2 11

3 $\dfrac{dy}{dx} = 7$

4 $\dfrac{dy}{dx} = 3x^2$

5 $\dfrac{dy}{dx} = 2x - 3$

6 $f'(x) = 6x^2 - 4$

7 $\dfrac{dy}{dx} = -2x^{-3}$ or $-\dfrac{2}{x^3}$

8 $\dfrac{dy}{dx} = \dfrac{1}{2}x^{-\frac{1}{2}} = \dfrac{1}{2\sqrt{x}}$

9 $\dfrac{dy}{dx} = 5x^4 - 1$

10

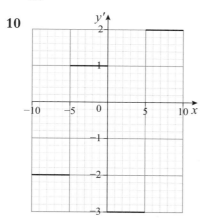

11

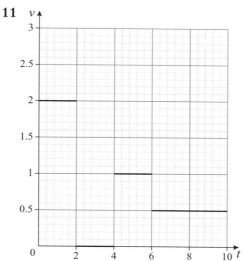

12 $-3 < x < -1$

13 $-1\dfrac{2}{3} < x < 1$

14 (i) 7

  (ii) 3.31

  (iii) 3.0301

  (iv) It is approaching 3.

15 Minimum value at the point.

16 $3x^2$

  This is a derivation question, so a short answer cannot be given. Go online to see the worked solution and mark scheme.

17 $2x - 4$

  This is a derivation question, so a short answer cannot be given. Go online to see the worked solution and mark scheme.

18 9

  This is a verify question, so a short answer cannot be given. Go online to see the worked solution and mark scheme.

19 $a = 6$, $b = 10$

20 (i) $(0, 0)$

This is a show-that question, so a short answer cannot be given. Go online to see the worked solution and mark scheme.

  (ii) This is a show-that question, so a short answer cannot be given. Go online to see the worked solution and mark scheme.

  (iii) $y = x$ is a tangent to $y = x^3 + x$.

21 (i) $\dfrac{dC}{dv} = 200v - \dfrac{200\,000}{v^2}$

  (ii) $v > 10$ knots

  (iii) 10 knots

22 (i) $y = 2$ and $y = -2$

  (ii) $x = 1$ and $x = -1$

  (iii) 8 square units

23 (i) Equation of tangent is $y = 10x - 14$.

Equation of normal is $y = -\dfrac{1}{10}x + \dfrac{31}{5}$.

  (ii) 182 square units

24 (i) This is a show-that question, so a short answer cannot be given. Go online to see the worked solution and mark scheme.

  (ii) $125\,\text{cm}^3$

25 (i) This is a show-that question, so a short answer cannot be given. Go online to see the worked solution and mark scheme.

  (ii) $A = 19.0$ (3 s.f.)

  (iii) $x = \dfrac{12}{11}\left(3 + \dfrac{\sqrt{3}}{2}\right), h = \dfrac{9}{11}\left(5 - \sqrt{3}\right)$

26 $349\,\text{cm}^2$ (3 s.f.)

11 Integration

1 $y = x^2 + c$

2 $y = x^3 - x^2 + c$

3 $f(x) = \dfrac{2}{3}x^{\frac{3}{2}} + c$

4 $y = \dfrac{5}{2}x^2 - 2x - 5$

5 $y = -\dfrac{4}{x} + 3x - 3$

6 20

7 $\dfrac{2}{5}x^{\frac{5}{2}} + c$

8 $s = 2t^3 - 2t^2$

9 $17\dfrac{1}{3}$

10 $\dfrac{x^4}{4} + \dfrac{1}{2x^2} + c$

11 8 square units

12 $(0, -6)$, $(1, -8)$ and $(-1, -8)$

13 $11\frac{5}{6}$ square units

14 $15\frac{5}{6}$

15 (i)

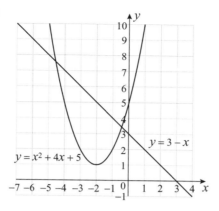

(ii) 11.7 square units (3 s.f.)

12 Vectors

1 $\begin{pmatrix} 7 \\ -10 \end{pmatrix}$

2 17

3 158°

4 $\begin{pmatrix} 6\sqrt{3} \\ -6 \end{pmatrix}$ or $6\sqrt{3}\,\mathbf{i} - 6\mathbf{j}$

5 $-\dfrac{2}{\sqrt{5}}\mathbf{i} + \dfrac{1}{\sqrt{5}}\mathbf{j}$

6 $\sqrt{52}$

7 (i) $-4\mathbf{i} + 3\mathbf{j}$

  (ii) $-0.8\mathbf{i} + 0.6\mathbf{j}$

  (iii) This is a show-that question, so a short answer cannot be given. Go online to see the worked solution and mark scheme.

8 $2\mathbf{i} + 3.60\mathbf{j}$

9 (i)

  (ii) $\begin{pmatrix} 5 \\ 0 \end{pmatrix}$

  (iii) $m = \dfrac{12}{7}, n = \dfrac{1}{14}$

10 (i) $\begin{pmatrix} 4 \\ -4 \end{pmatrix}$

  (ii) $2\sqrt{5}$
   117°

  (iii) $(3, -2)$

  (iv) $\left(3, -2\frac{2}{3}\right)$

13 Exponentials and logarithms

1   -4

2   $-\dfrac{2}{3}$

3   32

4   1

5   $\dfrac{1}{2}$

6   -1

7   $y = \log_a x$

8   2.10

9

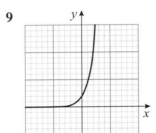

10

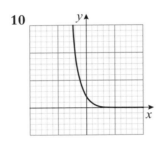

11 61.4

12 $n = 12$ is the largest value.

13 $4e^{4t}$

14 $\log_a \dfrac{y}{x}$

15 (i)   £2629

   (ii) $V = 1200 \times 1.04^t$

  (iii) $t = -4.65$ so 1996

16 (i) $x = \dfrac{1}{3}\ln\left(\dfrac{b-a}{5}\right)$

   (ii) $x = \dfrac{y}{(1-y)}$

17 (i)

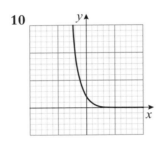

   (ii) $4\ln 2$

18 (i) 0.68 m and 4.51 m

(ii) As $t \to \infty$, $h \to \infty$, but the tree will not keep growing taller for ever.

(iii) $\ln 0$ is not defined.

(iv) 0.66 m and 4.57 m. Very similar to model A.

(v) $t \to \infty$, $e^{-0.1t} \to 0$, $h \to 4.6$ m

(vi) $h = -1.9 < 0$, but height cannot be negative.

(vii) $0.65e^{-0.1t}$, the rate of growth of the tree is 0.394 m per year.

(viii) 2.89 m and 3.72 m, so model A is better for this.

19 (i) $\ln N = \ln k + n \ln x$, so graph will be a straight line with gradient n and intercept $\ln k$.

(ii)

$\ln x$	−1.20	−0.69	−0.22	0	0.18	0.41
$\ln N$	3.30	3.04	2.77	2.64	2.56	2.48

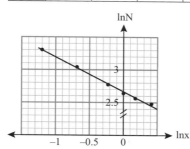

(iii) $k = 14.44$, $n = -0.5256$

(iv) 16.8 swings

(v) 0.93 m

20 (i) This is a show-that question, so a short answer cannot be given. Go online to see the worked solution and mark scheme.

(ii) It would be a straight line with gradient k and intercept $\ln A$.

(iii)

x	51	61	71	81	91
$\ln P$	3.3094	3.3645	3.4479	3.5535	3.6324

(iv)

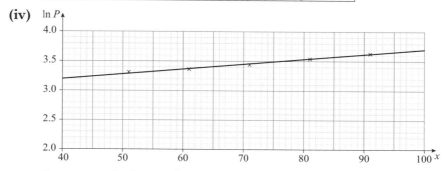

Scatter graph close to linear, so model is valid.

(v) $k = 0.00835$, $A = 17.62$ which is an estimate of population in millions of the UK in 1800.

(vi) 40.95 million, 62.17 million

(vii) 1901 $P = 40.95$ million people which is close to the real value.

1951 $P = 62.17$ million people which is not close to the real value, so it is not a good model for the twentieth century.

(viii) $\dfrac{dP}{dx} = 0.147127e^{0.00835x}$

0.3146 represents rate of growth of the population in 1891, at 0.3146 million people per year.

14 Data collection

1 Advantage: representative.
  Disadvantage: data need to be listed.

2 Every sample of a given size is equally likely to be selected.

3 Advantage: easy to collect.
  Disadvantage: likely to be biased.

4 It is expensive on resources (time and money) to survey the whole population compared with taking a sample.

5 All cups of 'coffee to go'.

6 (i) Gaps in the data; countries vary greatly in size, stage of development, etc.

  (ii) Clean the data, have a strategy for collecting another item; use a different sampling technique.

  (iii) Stratify by sub-region.

7 (i) Use the list as a sampling frame, generate a random number from 1 to the total number of countries and select that country until obtained total of 30 countries.

  (ii) Some data may be missing.
       Exclude that country, and generate another random number.

8 (i) A cluster sample chooses one or more representative groups of items and ignores the other groups.
       The method described does not choose a whole group.

  (ii) Quota sampling

9 He will miss out categories of shopper who don't use corner shops.
  He will include only shoppers who use corner shops on a Saturday.
  Some shoppers will not be prepared to fill out a questionnaire, however short.

10 7, 18, 25 or 8, 17, 25

15 Data processing, presentation and interpretation

1 'tr' means a trace of rain, so recording zero would be appropriate for an average.
  'n/a' means there were no data recorded, so these items should be excluded from the data.

2 Height, shoe size

3 A and B

4 C

5 Mean = 4, mode = 2, median = 4.5

6 There is no correlation as there are two sets of data here.

7 It is likely that Mundher has looked at the angle for the two groups, as one is 180° and the other is 90°.
  A pie chart shows the proportions of a whole rather than the actual amounts.

8 (i) Positive correlation

  (ii) 4 ± 1.

    As 4 falls within the range of the data we are interpolating, that is quite reliable.

9 Mean = 25.4

  Standard deviation = 8.61

10 Mean = 12.5

  Standard deviation = 2.729

  6 is an outlier.

11 (i) Pass mark is 60.

  (ii) 23 to 77 = 54

12

Mass (kg)	Frequency
$30 \leqslant m < 35$	$3 \times 0.2 \times 5 = 3$
$35 \leqslant m < 40$	4
$40 \leqslant m < 50$	14
$50 \leqslant m < 60$	$3 \times 0.2 \times 10 = 6$
$60 \leqslant m < 75$	3

Number above 55 kg = six snow leopards

13 (i) Median = 459.5

  Lower quartile (LQ) = 455,
  Upper quartile (UQ) = 474

  (ii)

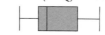

results of the more recent survey

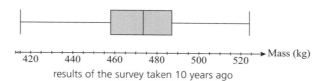

results of the survey taken 10 years ago

Or read off the values from the box and whisker plot.

Minimum = 415, lower quartile = 458, median = 473, upper quartile = 487, maximum = 524.

Each measure in the second survey is less than in the first survey (maximum is 484 compared with 524) except for the minimum value. The change is less marked for the items below the median, with the lower quartile being only slightly less (455 compared with 458). The data in the later survey are less variable than in the first one, with both range (48 compared with 109) and inter-quartile range (IQR) (19 compared with 29) being smaller.

  (iii) The decrease in the measures suggests that there is a decrease in mass of the polar bears: upper quartile is 474 compared with 487, median is 459.5 compared with 473.

    It is less pronounced at the lower end, with the lower quartile being only slightly less (455 compared with 458).

  (iv) There are no outliers.

    $LQ - 1.5 \times IQR = 458 - 43.5 = 414.5 < 415$, so none at the lower end.

    $UQ + 1.5 \times IQR = 487 + 43.5 = 530.5 > 524$, so none at the higher end.

14 Graph A: 30 throws, one of the numbers hasn't occurred, more likely with fewer throws.

Graph B: 100 throws, closest to theoretical distribution.

Graph C: 50 throws, all outcomes covered but less close to the theoretical distribution than B.

15 Much depends on the reason for Noah's report.

If he is considering the weather from day to day to evaluate the range of tourist attractions in the area and how many wet weather ones are needed then the first graph is more appropriate.

If he is looking at extreme weather, perhaps the capacity of the drains to deal with high rainfall, then the second graph is more appropriate.

16 Paper 1 was harder because the marks were lower.

Measure	Paper 1	Paper 2
Minimum	15	32
Lower quartile	35	46.5
Median	42	52
Upper quartile	49.5	57.5
Maximum	75	78

The graph shows the marks are lower, and the measures support this.

17 (i) Positive correlation

(ii) £1000 ± 50 by eye using likely line of best fit.

(iii) £100 ± 50 by going back along the likely line of best fit.

(iv) The estimate in part **(ii)** is likely to be better, as it is interpolated (within the range of the data), whereas the estimate in part **(iii)** is extrapolated (beyond the range of the data).

(v) Height does not affect salary as there is no mechanism for that to happen. Correlation does not mean causation. However, tall people may be seen to be more 'grown-up' than shorter people, and this subconscious perception may affect people's chances of acquiring more responsible jobs.

18 The mean for the Netherlands, Belgium and France = 81.4.

This is 1 year more than the average life expectancy in the UK.

The average life expectancy in the UK is greater than that for Belgium but less than those for the Netherlands and France.

19 Mean of Supplier B's eggs = 56.1 g

Standard deviation of Supplier B's eggs = 8.3361 = 8.3 g

Supplier A's eggs are larger, 61 > 56.1, but more variable as well, 10.2 > 8.3

Cupcakes need a consistent mixture, as the ratio of ingredients is important for the quality of the sponge, so she should use Supplier B.

20 (i)

Measure	Species A	Species B
Median	80.5	86.5
Lower quartile	72	81
Upper quartile	85	95
IQR	13	14
Range	41	47

Species B grows higher, as shown by higher median and quartiles.

Species A is more consistent, as shown by lower IQR and range.

Species A may be better for a border, as the plants are likely to be more similar in height.

(ii) Species A: no outliers.

Species B: no outliers.

16 Probability

1 $\frac{2}{6}$

2 $\frac{6}{36}$

3 (i) $\frac{1}{36}$

(ii) 5

4 0.6

5 10 different groups

6 0.984

7 $\frac{25}{72}$

8 Blue and green

9 (i) $\frac{47}{66}$

(ii) $\frac{15}{22}$

10 (i) $\frac{3}{5}$

(ii) $\frac{4}{5}$

(iii) $\frac{1}{5}$

11 (i) $\frac{1}{42}$

(ii) $\frac{10}{21}$

12 (i) Sample space diagram for $|X - Y|$

X: Y	1	2	3	4	5	6
1	0	1	2	3	4	5
2	1	0	1	2	3	4
3	2	1	0	1	2	3
4	3	2	1	0	1	2
5	4	3	2	1	0	1
6	5	4	3	2	1	0

Probability distribution for $|X - Y|$

r	0	1	2	3	4	5
$P(|X - Y| = r)$	$\frac{6}{36}$	$\frac{10}{36}$	$\frac{8}{36}$	$\frac{6}{36}$	$\frac{4}{36}$	$\frac{2}{36}$

(ii) 50

13 This is a show-that question, so a short answer cannot be given. Go online to see the worked solution and mark scheme.

17 The binomial distribution

1 $X \sim B(20, 0.65)$

2 0.402

3 $X \sim B(60, 0.35)$

4 30

5 $Y \sim B(100, \frac{1}{36})$

6 3.4×10^{-7}

7 (i) Binomial distribution with $n = 50$ and $p = 0.15$

(ii) 0.209

(iii) 7.5

(iv) $P(X = 6) = 0.1419$, $P(X = 7) = 0.1575$,
$P(X = 8) = 0.1493$ so 7 is the most likely number.

8 (i) Each throw equally likely to succeed, independent trials.

(ii) 0.3292, 0.1646, 0.0412

(iii) 0.2629

(iv) 0.3686

9 (i) 0.328

(ii) 0.168

(iii) 0.302

(iv) The probability of being stuck twice in one week and not in the other has not been included.

10 (i) $p = 0.0174$ (evidence of trials, e.g. $p = 0.01735$ and $p = 0.01745$, need to be seen).

(ii) 1.3275×10^{-6} which is extremely unlikely.

(iii) Breakages are not independent − several can occur at once.
Or, eggs in the middle of the box may not have the same probability of breakages as others.

11 (i) $k = \dfrac{1}{30}$

(ii) $\dfrac{59}{150}$

18 Statistical hypothesis testing using the binomial distribution

1 $H_0 : p = 0.5$

2 $H_0 : p = \dfrac{1}{6}, H_1 : p > \dfrac{1}{6}$

3 $H_1 : p \neq 0.7$

4 $\{13, 14, 15\}$

5 Reject the null hypothesis/accept the alternative hypothesis.

6 $\{0, 1, 9, 10\}$

7 (i) $H_0 : p = 0.2, H_1 : p < 0.2$, looking for decrease (one-tailed).

(ii) $P(X \leqslant 1) = 0.0692 < 10\%$ so significant. Reject H_0.
Enough evidence to suggest that the probability has decreased from 20%.

8 (i) $H_0 : p = 0.3, H_1 : p \neq 0.3$, looking for any difference (two-tailed).

(ii) $X \in \{0, 1, 2, 13, 14 \dots 24\}$

(iii) $X = 3$ is not in critical region so not significant.
Accept H_0.
Not enough evidence at 5% level to suggest any change.

9 (i) Each child has same probability of voting for Winston.
They vote independently.

(ii) $H_0 : p = 0.5, H_1 : p \neq 0.5$, looking for any change (two-tailed).

$P(X \geqslant 20) = 0.0494 > 2\frac{1}{2}\%$ or critical region $X \in \{0 \dots 9, 21 \dots 30\}$
X not in critical region.

Not significant.

Accept H_0.

Not enough evidence at 5% level to suggest any change in voting.

10 (i) $H_0 : p = 0.08, H_1 : p < 0.08$, looking for decrease to improve the situation (one-tailed).

(ii) $X \in \{0, 1\}$

(iii) $X = 1$ is in the critical region so significant. Reject H_0.
There is enough evidence at 10% level to suggest an improvement in the situation – the proportion of cars going much too fast has decreased.

19 Kinematics

1 Speed

2 Acceleration

3 Distance travelled

4 Distance travelled is a scalar quantity, so the direction does not matter.
Displacement is a vector quantity, so the direction is significant.

5

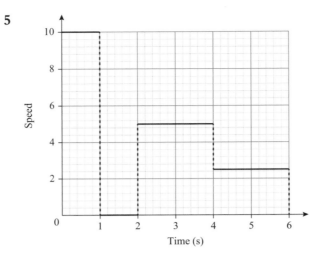

6

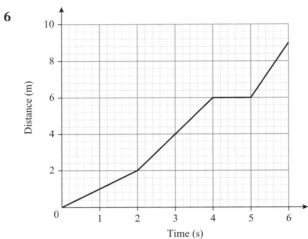

7 $44\,\text{m}\,\text{s}^{-1}$

8 $8\,\text{m}$

9 $u = 5.74\,\text{m}\,\text{s}^{-1}$

10 $t = 4\,\text{s}$

11 $6.88\,\text{m}\,\text{s}^{-1}$

12 $t = 2.02\,\text{s}$

13 $123\,\text{m}$

14 $u = -2.5\,\text{m}\,\text{s}^{-1}$

15 (i) Between $t = 4$ and $t = 10\,\text{s}$

   (ii) $6\,\text{s}$

   (iii) $0.8\,\text{m}\,\text{s}^{-1}$

16 $18\,000\,\text{m}$

17 $0.5\,\text{m}\,\text{s}^{-1}$
   $-0.05\,\text{m}\,\text{s}^{-2}$

18 (i) $6.25\,\text{m}\,\text{s}^{-2}$

   (ii) $15\,\text{m}\,\text{s}^{-1}$

19 (i) $81.6\,\text{m}$

   (ii) $40\,\text{m}\,\text{s}^{-1}$

20 (i) $60\,\text{s}$

   (ii) $1440\,\text{m}$

21 (i) $-2\,\mathrm{m\,s^{-2}}$

  (ii) $6.5\,\mathrm{s}$

  (iii) $18\,\mathrm{m}$

22 $49\,\mathrm{m\,s^{-1}}$

23 (i)

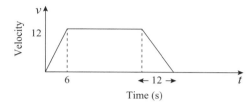

  (ii) $84\,\mathrm{s}$

  (iii) $15.8\,\mathrm{s}$

20 Forces and Newton's laws of motion

1 Kilograms, metres, seconds

2

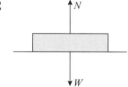

3 $2\mathbf{i} - 2\mathbf{j}$

4 $78.4\,\mathrm{N}$

5 $0.4\,\mathrm{m\,s^{-2}}$

6 $12\,\mathrm{N}$

7 $40\,\mathrm{kg}$

8

9 $-1.6\mathbf{i} + 4\mathbf{j}$

10 $5\,\mathrm{m\,s^{-2}}$

11

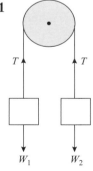

12 $5\,\mathrm{kg}\colon T - 5g = 5a$

  $7\,\mathrm{kg}\colon 7g - T = 7a$

13 $a = \dfrac{g}{7}\quad T = \dfrac{24}{7}g$

14 $a = \frac{1}{7}$

$T = 614\,\text{N}$

15 This is a show-that question, so a short answer cannot be given.
Go online to see the worked solution and mark scheme.

16 Either there is no acceleration or the towbar is assumed to have negligible mass.

17 $\mathbf{i} - 19\mathbf{j}\,\text{N}$

18 (i) $600\,\text{N}$

(ii) Less than $1\,\text{m}\,\text{s}^{-2}$

19 $1.43\,\text{s}$

20 (i) $a = \frac{g}{9}\,\text{m}\,\text{s}^{-2}$

(ii) $F = \frac{17g}{9}\,\text{N}$

21 (i) $a = 0.518\,\text{m}\,\text{s}^{-2}$

(ii) $m = 0.969\,\text{kg}$

22 $T_{PA} = 502\,\text{N}$

$T_{AB} = 312\,\text{N}$

$m = 72.7\,\text{kg}$

23 Maximum of five passengers

21 Variable acceleration

1 $15t^2 - 2t$

2 $-4t$

3 Velocity $= \frac{1}{4}t^2 - 5t$

4 $\frac{1}{30}t^3 - t^2$

5 $4t - \frac{t^2}{2} + 3$

6 $476\,\text{m}$

7 $\frac{t^3}{2} - \frac{5}{2}t^2 + 3$

8 This is a show-that question, so a short answer cannot be given. Go online to see the worked solution and mark scheme.

9 (i) $-290\,\text{m}\,\text{s}^{-1}$

(ii) $20\,\text{s}$

10 (i) This is a show-that question, so a short answer cannot be given. Go online to see the worked solution and mark scheme.

(ii) $8\,\text{m}$

11

12 (i) $s = 2t^3 - 2t^2 - 8t + 8$

(ii) $1\,\text{s}$ and $2\,\text{s}$